A MODERN Horse Herbal

A MODERN Horse Herbal

HILARY PAGE SELF

VETERINARY ADVISER
TIM COUZENS
BVetMed, MRCVS, VetMFHom

KENILWORTH PRESS

AUTHOR'S NOTE

All the information contained in this book stems from the author's personal experience of using herbs on her own horses, and on those belonging to other people, with veterinary supervision.

First published in Great Britain 1996 by
Kenilworth Press
Addington
Buckingham
MK18 2JR

© Hilary Page Self 1996

All rights reserved. No part of this publication may be reproduced, stored in a retrieval system, or transmitted in any form or by any means, electronic, mechanical, photocopying, recording or otherwise, without the prior permission of the copyright holder.

British Library Cataloguing in Publication Data
A CIP record for this book is available from the British Library

ISBN 1-872082-85-8

Typeset in 11/13 Bembo
Design by Paul Saunders
Illustrations by Carole Vincer
Typesetting and layout by Kenilworth Press
Printed and bound in Great Britain by
WBC Book Manufacturers Ltd, Bridgend

DISCLAIMER

This book is not to be used in place of veterinary care and expertise. No responsibility can be accepted by the author, publishers or distributors of this book for the application of any of the enclosed information in practice.

CONTENTS

Foreword *by Tim Couzens, BVetMed, MRCVS, VetMFHom*   7
Acknowledgements   8
Introduction   9
How to Use This Book   12
Choosing a Remedy   14
Dosages for Herbs   15
How to Prepare and Use Herbs   17
  • Dry or Fresh-cut Herbal Preparations • Infusions/Teas/Brews
  • Decoctions • Tinctures/Extracts/Concentrates • Compresses
  • Poultices • Ointments

PART ONE
Materia Medica   25
Non-Herbal Miscellany   64
Alternative Antibiotics   69

PART TWO
Common Ailments   73

PART THREE
Alternative Therapies   143
  • Bach Flower Remedies • Aromatherapy and Essential Oils
  • Homoeopathy *by Tim Couzens, BVetMed, MRCVS, VetMFHom*
  • Radionics • Chiropractic • Other Therapies and Treatments

Glossary   154
References   158
Bibliography and Further Reading   160
Useful Addresses   164
  • Recommended Suppliers • Alternative and
    Complementary Therapies
Index   169

To Rags and Ryan – the beginning

FOREWORD

There is a unique link between herbs, healing and health. For this simple reason plant-based medicines have a traditional association with most world cultures. Much of the knowledge is laid down as folklore or documented in herbals, classical books detailing the medicinal properties of common plants used in herbal medicine.

Not surprisingly, frequent mention is made of treating domesticated animals, a natural step in view of the abundance of healing plants in our hedgerows. In the past the veterinary profession made good use of this knowledge, particularly in treating horses. Until the earlier part of this century, herbal remedies, including many common to modern herbalism, such as marshmallow and liquorice, formed the mainstay of equine veterinary treatment. This tradition continued until advances in modern medicine began to relegate herbal remedies to a state of obscurity and ridicule.

Today the art of using herbal remedies to treat horses is undergoing a triumphant revival, heralded by the growing interest in complementary therapies. This book is a celebration of those skills and symbolises the relationship between man, Nature's pharmacy and the horse. It reflects the light of modern herbal medicine and seeks to re-establish links with the traditional herbal skills of the past.

In an age of advanced medical technology, this book serves as a timely reminder that Nature holds remarkable healing powers and can provide a simple solution to many of the health problems seen with increasing regularity in horses today.

Tim Couzens
BVetMed, MRCVS, VetMFHom

Acknowledgements

First and foremost my thanks go to all the animals I have had the pleasure and privilege to care for – you have been my inspiration and my reward.

To Tony, my husband – 'You can do it, Hils' – thanks for all your support and for giving me the time off to write this book.

Thanks to my family – Anita, Clive and Liana – for your enthusiasm, encouragement and, most of all, for believing in me. In particular, thank you, Mum, for your patient proof-reading.

To Tim Couzens, for writing the excellent section on homoeopathy, for checking the veterinary details, and for writing such a lovely foreword.

To Annie Dent, who provided the informative section on aromatherapy along with the author's photo – thanks, Annie, for your friendship and advice.

To Ian Mole of Equine Marketing, for providing me with the excellent 'user friendly' section on probiotics.

My very grateful thanks to Helen Leather for her permission to use the 'Cracker' story, and to registered aromatherapist Caroline Ingraham for checking the aromatherapy details.

Thanks to William and Katrina Prull (the American Herbies) for helping me source the US research information and advising on US terminology.

Finally, a big thank you to Lesley Gowers of Kenilworth Press, who made me put thoughts into action – your patience and enthusiasm have helped make this a very enjoyable period of my life.

Hilary Page Self
July 1996

INTRODUCTION

'Our ancestors discovered the rudiments of their folk medicine in the healing plants sought out by animals suffering from alimentary disturbances, fever and wounds. By observing how animals cure themselves from disease, they learned how to keep themselves healthy by Nature's own methods.'

Dr D.C. Jarvis *(Folk Medicine)*

Twentieth-century man has lost the sixth sense which allows him to know instinctively what is good for him. Mercifully, animals have not, and if returned to a more natural environment could fend for themselves very successfully. They would not be plagued with the ailments which domesticity, intensive farming and man's greed have created. Unfortunately the majority of our farm animals do not get the opportunity to exercise their instinctive knowledge, being kept in what can only be described as 'unnatural conditions', with little or no control over what they eat. Today, many farm animals such as cattle, pigs and hens are confined to housing which allows them little or no access to grassland and very often no natural sunlight; they are forced to stand on concrete floors and are crated in pens which restrict their movements. The resulting poor physical health, not to mention mental distress, is then treated with a cocktail of chemicals that subsequently enter the food chain, with all the future consequences that may entail.

Horses have fared better than the majority of farm animals, in that the place they fill in our lives is generally for pleasure rather than profit. They have retained the instinct to know what is good for them, but are often prevented from using it by being restricted at best to paddocks devoid of natural herbage, and at worse confined to stables for the majority of their working lives. One only has to see how well the British native ponies cope with hardship to appreciate how much goodness they take from the harsh landscapes in which they live. It is only when they are removed from these sparse conditions and introduced to herb-free, protein-rich paddocks that conditions such as laminitis and sweet itch arise. Long-term confinement in stables (for sweet-itch sufferers) or in 'starvation paddocks' (for laminitic ponies), the common solutions for these conditions, subsequently create their own set of problems.

The role of the horse has changed dramatically in the last fifty years. Through selective breeding, the 'sports horse' has been created. This animal now runs faster, and jumps higher and longer than his forbears. He is comparable with the modern-day athletes that we see breaking world records on race tracks around the world; unfortunately he is just as prone to injury!

Even if the horse is not being used in strenuous competition, he still has to cope with mental and physical pressures. Poor air quality and the breakdown of the ozone layer, which results in an increase in harmful ultra violet rays in the atmosphere, affects us all and does not leave the horse unscathed. Allergies, photosensitivity and respiratory conditions are now far more commonplace, resulting in a greater use of cortisone and other steroidal-based drugs. A series of studies carried out on racehorses over the past ten years both in America and Hong Kong have shown that as many as 80% of the horses examined, whilst in training, were found to be suffering from gastric ulceration.[1] Although no singular cause for this ulceration was found during the studies, it was generally concluded that stress, insufficient fibre, incorrect feeding programmes, and the use of non-steroidal anti-inflammatory drugs may contribute to the incidence of gastric ulceration. Interestingly, in the Hong Kong studies the incidence of ulceration drops to 50% in horses who have retired from training for a month or more. I leave you to draw your own conclusions.

Science now plays a greater role in the control of our animal feeds and the drugs that we use to enhance their performance. Stresses placed on our horses, not only by their training but also by the way in which they are kept, have produced new problems which did not exist twenty years ago. Unfortunately it has also produced an even greater reliance on the 'quick fix' approach.

One cannot deny the benefits of conventional medicine, particularly in life-threatening conditions, but now, more than ever, there is a need to look back to the benefits that herbs can have on the animals in our care. Herbal medicine can work side by side with conventional medicines and has often effected a cure when all other avenues have failed. Between 1900 and 1935 veterinary surgeons used herbal remedies extensively and successfully, as part of their treatment; this is well documented.[2] It is only in more recent years that we have come to rely more heavily on synthetically produced drugs. Throughout the ages herbal remedies have been used effectively by generations of horsemen and women who knew the benefit of herbal medicine, both for themselves and their stock. I feel

INTRODUCTION

sure that it is by moving away from the use of these herbs and their gentle healing action that we have served our animals poorly. Perhaps, with a greater understanding of the benefits that herbs can bring to horses, we can start to redress the balance.

Lay people using herbs on their animals are often the butt of both derision and criticism from conventional practitioners and horse 'experts', who state that such remedies are unproven by trials and rely heavily on the human applications of herbs. To these people I say, 'Look to the past!', where observation of animals guided humans in their choice of plant remedies. A few months' trials in a modern laboratory, carried out in unnatural conditions, can scarcely compare with centuries of study, documented evidence, and successful practical application on both humans and animals, from such distinguished herbalists as Messrs Dioscorides and Galen (who wrote herbals in AD1 and AD129 respectively, Gerard (1545-1612), Culpeper (1616-1654) *et al*! Their findings can truly be said to have stood the test of time; moreover they resulted in the production of the very pharmacopoeias from which modern drugs have been synthesised.

How To Use This Book

In **Part One** you will find a Materia Medica of all the plants mentioned as remedies, their full botanical names to ensure precise identification, the parts used, their habitat, actions and uses, and suggested dosages. Following the Materia Medica are sections containing tips on using a number of non-herbal substances which can also benefit your horse's health, for example cider vinegar and honey, and advice on herbal antibiotics.

Notes on how to prepare and use the herbs, along with general guidance on dosages and how to choose a remedy, can be found conveniently at the front of the book on pages 14-21.

In **Part Two** I have taken the most common complaints experienced by horse owners in the 1990s and outlined the herbal treatments that can be administered. This part also includes ideas for dealing with a number of management problems.

When used as described, the herbs recommended as remedies should not interfere with, nor reduce the effectiveness of any conventional drugs/medication that the horse may be given. The majority of the herbs recommended are either native to or readily available throughout Europe and the USA. All of the herbs suggested are on the UK's General Sales List (a list produced by the Medicines Commission, the governing body responsible for issuing licenses for the sale of all medicines) or are recognised foods, and they are neither poisonous nor toxic. Many of them are readily available by picking fresh (but do ensure that all plants are correctly identified and gather them away from areas contaminated either by chemical spraying or traffic pollution). If this is not feasible or time does not permit, all the herbs covered in this book can be purchased from health food stores or herbal suppliers (see the list of recommended suppliers in the Useful Addresses section at the back of the book).

Part Three gives brief descriptions of a number of alternative therapies which can be used alongside conventional and/or herbal medicine, together with some interesting case histories.

CAUTION

It must be mentioned here that this book is not to be used in place of veterinary care. If the horse is unwell, then the owner must call a veterinary surgeon to attend to the animal. If you wish to give your own horse a herbal remedy, you should do so under the supervision of your vet – if you fail to do this and your horse is suffering from a serious condition, you could be breaking the law. In the UK the Veterinary Surgeons Act (1966) makes it illegal for any person who is not a veterinary surgeon to treat or advise on the treatment of any condition or disease of a horse. American laws regarding these matters are generally more relaxed, but readers outside the UK are advised to consult local sources for current regulations governing animal treatment and the use of herbal remedies.

Choosing a Remedy

It has been my experience over a number of years that horses are extremely sensitive to what is good for them. Normally, unless forced by starvation, they will not voluntarily eat any fodder which will do them harm. Therefore be aware of this, and be guided by it. If the horse is not enthusiastic about a particular herb, find an alternative one to feed. Horses are like humans and have allergies and preferences; the same things do not suit them all. Take note of what the horse seems to be searching for and try to understand what he/she is trying to tell you.

Another thing that I have found is that the horse will 'tell you' once it has had enough of a particular course of treatment. Many horses once 'cured' will start to refuse the very herb that brought about that 'cure'. This is just another example of the horse exercising its instinctive knowledge of what it needs. So once again, listen to your horse.

DOSAGES FOR HERBS

Herbs work by treating the body as a whole, as opposed to treating the symptoms in isolation. It is therefore necessary to remember that any herbal remedy will take some time to be absorbed by the system, and for any improvement to be seen.

> **IMPORTANT**
>
> **Unlike synthetically produced medicines, the herbs featured in the Materia Medica which follows do not require absolute accuracy of dosage to guarantee success or safety.**

The recommended dosage (normally in handfuls or grams) is made on the assumption that one herb only is being given to a 15.2-16hh horse. Remember, though, that horses, like humans, all respond differently and in many cases the size of the horse does not necessarily dictate the quantity. I have had experience of some 17.2hh horses who, because they were particularly responsive to herbal treatment, needed a fraction of the dosage I would normally give a small pony.

Unlike drugs, where 10cc can be the difference between life and death, herbs are not an exact science and, when using the herbs detailed in this book, being 10 or 20 grams out on the odd occasion will not make a great deal of difference.

However, for readers who feel they need slightly more accurate guidance, an average size pony 13.2hh will need approximately 25-30 grams of dried herbs per day, whereas a large horse 16.1hh and upwards will need between 30 and 50 grams. Should a combination of several herbs be indicated, then a general rule of thumb is to divide the total dosage or weight in grams between the number of herbs used – i.e. if a combination of three herbs is being given to a 13.2hh pony, then use 7-

10 grams of each herb, and so on.

I have generally found with horses that it is neither necessary nor very often convenient to produce infusions or tinctures for ingestion. The majority of horses and ponies will readily take the herbs in their natural form, either on their own or mixed with their normal daily feed.

Infusions and tinctures do tend to concentrate the potency of the herb, and can make the constituents of the plants more bio-available. Therefore if the horse is off his food, or time is of the essence, then an infusion can be made and administered quickly via an oral syringe. (See the next chapter on preparation.)

How To Prepare and Use Herbs

Dry or Fresh-cut Herbal Preparations

This, as far as horses and horse owners are concerned, is the easiest and least complicated way of feeding herbs, particularly those in leaf and flower form. The horse's gut is designed to break down cellulose and fibre and horses respond well to herbs fed in this way.

The herbs, be they dried or fresh, can be fed either on their own or mixed in the horse's daily ration. Remember to chop them up prior to feeding, as this will start to break down the cell walls of the plants, which will help with digestion and absorption. Some herbs, such as burdock, nettle and comfrey, may need to be bruised or allowed to wilt if fed in their fresh form, as they can be prickly or hairy, which some horses don't like.

The other benefit of feeding the herbs in this way is that the horse will taste them! This may seem a bit obvious but the bitter herbs such as Horehound and Wormwood work by the bitterness in the mouth stimulating the nervous system to promote appetite and the production of digestive juices.

This is also a good way of assessing what the horse does and doesn't want. As I have said earlier, be guided by your horse's instincts. The horse usually knows what is needed, so if necessary find an alternative.

Infusions – Teas – Brews

The next two methods of preparing herbs – infusions and decoctions – are a little more time-consuming; however, they are a better method to use if a quicker response is needed. By adding boiling water, the active constituents of the herbs are made more readily available and are more easily absorbed by the horse. However, it must be pointed out that herbal treatments rarely produce instantaneous results; owners must be patient and several weeks can elapse before an improvement is seen.

Infusions, teas and brews all describe the same thing – namely the method of making a tea out of the herbs. This is a simple way to prepare herbs for horses and the liquid can be fed neat as a drink (sometimes sweetening the tea with honey – preferably unpasteurised – brown sugar or liquorice root can encourage the horse to drink), syringed down the throat, added to feed or to a bran mash, and if applicable used externally on bandages (comfrey is particularly effective when used in this way) and compresses, or as a rub for skin conditions.

In general the average dose for an infusion is: 1 good handful (approx 30 grams or 1 oz) of the leaves, green stems, flowers, seeds, etc., brewed in 1/2 litre/1 pint of boiling water. Double the quantity of herbs if you are using fresh herbs as they contain a higher water content.

Prepare the infusion in much the same way as you would make a pot of tea: pour the boiling water onto the herbs and allow them to steep for 15 minutes. The infusion can be used and fed warm if required, or allowed to cool. You can make larger quantities to provide you with enough for two to three days (no longer), but remember to store the infusion in a stoppered bottle in the refrigerator as it will ferment, especially in hot weather.

Dose: Give the horse 1/4 of a litre or 1/2 a pint of the infusion twice a day.

CAUTION

Never use any infusion which shows signs of fermentation, bubbling, or has an alcoholic smell – it could be harmful.

Decoctions

If you are using hard roots, bark, rhizomes, or some of the hard seeds, it is better to make a decoction rather than an infusion. The cell walls of the hard, woody parts of plants will not break down just by steeping them in boiling water.

To make a decoction put 1 good handful (30 grams or 1 oz) of the dried root, bark, etc., (or double the quantity of fresh roots) into a saucepan. Make sure the material is cut or broken into small pieces. Use a ceramic, earthenware or enamelled saucepan. Do not use aluminium pots. Add 3/4 litre or 1 1/2 pints of cold water. (These larger measures are to compensate for evaporation; the decoction should end up as approximately ½ litre or 1 pint of liquid.) Bring the water slowly to the boil, cover with a lid and simmer gently for 15 minutes. Strain the liquid whilst it is still hot.

I tend to give the softened roots etc. in the feed in addition to the liquid as the horse may extract further goodness from the material during digestion. Larger quantities of decoctions can be made and should be stored in the same way as infusions.

DOSE: Give the horse ¼ of a litre or ½ a pint twice a day. In addition, a dessertspoon of the freshly boiled and cooled softened bark, root or seed can be given twice daily.

CAUTION
Never use any decoction which shows signs of fermentation, bubbling, or has an alcoholic smell – it could be harmful.

AUTHOR'S NOTE
I do appreciate that many horse owners do not keep their horses at home and do not have facilities for preparing herbal teas etc. at their horse's yard. Nor it is always possible to rely on other people to carry out the necessary preparations. So for horses stabled away from home, using fresh or cut dried herbs is definitely easier, although one must expect the results to be a little slower.

Tinctures – Extracts – Concentrates

These preparations are a much more concentrated form of herbs and are administered in much smaller quantities. They can be made at home, using alcohol, a mixture of alcohol and water or glycerine, but they involve considerably more time and preparation. Assuming that most horse owners are usually desperately short of time, I have chosen not to go into the details of how these medicinal preparations are made. There are, however, a number of excellent herbals which do give detailed instructions on the preparation of tinctures etc., and several of these are listed at the back of the book. (*See* Bibliography and Further Reading.)

If tinctures and extracts etc. are required it is quicker, easier and often cheaper to buy the ready-made product from a reputable supplier. (*See* Recommended Suppliers, page 164.)

For dosages, follow the specific recommendations given in the ailments section.

Compresses

A compress, or fomentation as they are sometimes called, is one of the best ways of applying herbs to the skin. It is usually employed to speed up the healing process. A compress should be made of a clean cotton, linen, gauze or cotton wool pad, soaked in a hot infusion. Place the compress, as hot as possible (be careful that it is not too hot) onto the affected area, cover with a piece of plastic to keep the moisture in and bandage in place. The compress is most effective whilst it is still warm, and as the compress cools it should be replaced with a new warm one whenever possible. Vulnerary herbs such as calendula and hypericum are particularly good when used in compress form, as are stimulatory and diaphoretic herbs such as yarrow, vervain, thyme, and rosemary.

Poultices

A poultice is made by using the solid plant material rather than a liquid extract. You can use either fresh or dried herbs. If using fresh herbs the plant can often be applied directly to the affected area after first bruising the leaves. An excellent example of this is when using comfrey leaves: these should be pulped, spread onto a piece of cotton or linen and then applied directly to bruises or soft swellings and held in place with a bandage.

For dried herbs it is better to mix them with a 'medium' such as powdered slippery elm or powdered marshmallow root. Both of these have excellent 'drawing' properties and when mixed with liquids become slimy, soft and malleable. These 'mediums' will not only hold the herbs but enable you to mould the poultice to the contours of the horse's body, which I have found invaluable, as it is well known that horses always damage themselves in those really hard to reach places! If slippery elm or marshmallow are not available, then bran can be substituted, but it is better to wrap the bran in a flannel or piece of cotton prior to application, making it less messy when applied to the skin.

To make a poultice, use equal amounts of the chosen herb and the 'medium', add boiling water or hot apple cider vinegar and mix to a paste. Spread the paste onto a piece of cotton, linen, or cotton wool, cover the back of the material with a piece of plastic and bandage into position. Apply the poultice as hot as possible, taking care that it is not too hot! Repeat when the poultice has cooled. Keep an eye on horses that are known to have particularly sensitive skin, and leave the area unpoulticed periodically if the skin is becoming weakened. Poultices can be effective in helping to draw poison, and for puncture wounds. Always make up fresh poultices; never re-use an old one which may have become infected.

I was given a recipe for a poultice by an Indian doctor who suggested that I use equal parts of powdered horseradish, turmeric, and ginger to help reduce a non-malignant growth which our horse had on his jawline. It was the size of a tennis ball and he had first developed it at the age of six; he was fifteen when we bought him and the growth by then was causing him discomfort and severely restricting his ability to flex.

I made the poultice using castor oil to bind the herbs together and then held it in place with a pair of tights. It was in a very awkward area! The horse's coat, which is grey, turned yellow in that area due to the natural vegetable dye in the turmeric, but over a period of two weeks' poulticing twice a day, the soft tissue around the hard interior slowly dispersed and the size of the growth reduced by over 50%, allowing him to flex properly for the first time in many years. The horse is twenty-three now and still has full flexion. Needless to say, the yellow staining faded over a period of a few weeks. I have since used this poultice on both hard and stubborn soft swellings with excellent results.

> **CAUTION**
>
> **Care must be taken when using this recipe as it is easy to imagine the heat generated by this combination of herbs. The poultice should not be left on for long periods of time and an 8-hour gap must be allowed between poulticing. Horses with sensitive skins must be observed closely for blistering.**

Ointments

Most ointments used nowadays are prepared commercially, and there is such an extensive choice that unless you are very keen to produce your own it is easier and quicker to buy them ready made. If possible use ointments made with organic plant-based ingredients. Obviously some non-plant materials such as beeswax are not only acceptable but positively beneficial – honey, for example, has traditionally been used for healing purposes and is said to encourage hair growth. Oils such as comfrey, garlic and hypericum are used extensively in healing ointments. There is now also available a huge range of human homoeopathic ointments containing both plant extracts and essential oils, which can be used equally well on animals. (See list of suppliers at the end of the book.)

Part One

Materia Medica
Non-Herbal Miscellany
Alternative Antibiotics

Materia Medica

AGNUS CASTUS *Vitex agnus-castus*

COMMON NAMES: Chaste Tree, Chasteberry, Monk's Pepper.
HABITAT: Northern Mediterranean.
PARTS USED: Seeds/fruit.
COLLECTION: Autumn.
ACTIONS: Anaphrodisiac, hormonal normaliser.
USE: Agnus castus has been used and recommended since the fourth century BC for helping women to regulate their menstrual cycle, increase milk production and balance their hormones during the 'change of life'. It is reputed to be an anaphrodisiac and was laid at the feet of novices as they entered the monastery, hence its name Monk's Pepper. In more recent years it has undergone extensive trials in Germany and England and has been used with great success on women for both PMT (PMS) and menopausal problems.

How can this help horses I hear you ask? In 1991 I was introduced to a horse owner, who was having severe problems with her Dutch-bred show-jumping mare. The mare was extremely disturbed and unless tied up would damage herself by swinging round and biting at her flanks and stifles. The mare would do this both in the stable and in the field and was difficult to handle. The horse had been on a synthetic hormone-based drug for some length of time with limited success. This drug could not be given repeatedly and so the owner was looking for an alternative.

It seemed to me that the mare was displaying the sort of behaviour often shown by a woman with PMT – she was tense and irritable, and seemed 'at odds' with herself. She normally possessed the sweetest nature and was obviously not in control of her moods.

I had been using agnus castus on myself with great success, and felt that it was worth trying on the mare. With the vet's consent we gave her 15 grams of agnus castus daily, and in under twenty-one days her temperament and behaviour had improved dramatically. She no longer bit herself, was calmer and easier to handle, and because she was more comfortable she was able to round over the jumps more easily.

This is obviously not the answer to all hormonal problems in mares, but since then this herb has been used on a number of mares who have been diagnosed as having 'hormonal imbalances', and it has worked as an alternative to implants and drugs. Very often only a short course of the herb is necessary to restore the hormones to their correct balance. I have received many calls from mare owners desperate to try and sort out these problems. Some mares can become dangerous and unpredictable just before they come into season, and many owners have been given the option of either destroying the horse or putting her in foal to try and resolve the matter.

I have to say that in my experience breeding from these mares is not the answer. It can encourage breeding from unsuitable stock, and, to add insult to injury, the mental and physical problems that disappear during the pregnancy can reappear after the foal has been weaned!

DOSE: 15 grams daily in feed.

CAUTION

Agnus castus has no indications for use during pregnancy. Therefore if the intention is to breed from the mare it would be wise to discontinue the use of the herb during pregnancy. It can, however, be used prior to stud to help regulate the seasons if this is a problem.

ANISEED *Pimpinella anisum*

COMMON NAME: Anise.
HABITAT: Cultivated extensively in warmer climates, originally from Egypt.
PARTS USED: Dried seeds.
COLLECTION: The ripe, dried seeds should be collected from mid to late summer.
ACTIONS: Expectorant, antispasmodic, carminative, oestrogenic, parasiticide.
USE: Externally – the volatile oil can be used for parasitic infestations such as lice and scabies. Internally – the seeds are ideal as a carminative, for any digestive problems such as colic, and for persistent irritable coughing.
DOSE: Average dose is one handful of seeds daily.

ARNICA *Arnica montana*

COMMON NAMES: Leopard's Bane, European Arnica, Mountain Daisy.
HABITAT: Native to Europe in mountainous areas – not Britain. Cultivated in northern India.
PARTS USED: Dried flowers.
COLLECTION: Collect flowers in mid to late summer.
ACTIONS: Anti-inflammatory, stimulant.
USE: For bruising, wounds, shock, muscle pain, reducing fevers. Arnica has been the subject of recent studies and has been found to be an immuno-stimulant in that it appears to increase the resistance of animals to bacterial infections, by stimulating the action of white blood cells.[3, 4]

> **CAUTION**
>
> **Arnica should not be used internally in herbal or tincture form.** Only commercially produced homoeopathic preparations are entirely safe for internal use. Consult a qualified veterinary homoeopath with regard to dosage instructions.[5]

Arnica tincture, or creams and balms containing arnica, can be used externally, but avoid broken or sensitive skin, as irritation may occur. Use on any area where there is bruising, inflammation or rheumatic pain.

Arnica tincture can be added to water or distilled witch hazel for use as a general wash-down lotion, for tired, bruised muscles and legs. Arnica is ideal for use in the wash-down water used for cooling competition horses (such as during endurance rides or in the 10-minute box in eventing etc.). Add 5ml of arnica tincture to 2 litres/3½ pints of water.

BONESET *Eupatorium perfoliatum*

COMMON NAMES: Feverwort, Thoroughwort.
HABITAT: North-eastern USA, in damp/wet pastures.
PARTS USED: Dried aerial parts.
COLLECTION: In late summer/early autumn or as soon as the flowers open.
ACTIONS: Vasodilator, diaphoretic, gentle laxative, antispasmodic, immuno-stimulatory.
USE: For any sort of fever control. Particularly good for influenza symptoms and especially for bone pain. For upper respiratory tract mucous congestion, and constipation.
DOSE: 1 handful of the dried herb daily.

BUCHU *Barosma betulina*

COMMON NAMES: Short, Oval or Long Buchu.
HABITAT: South Africa.
PARTS USED: Leaves.
COLLECTION: Collect the leaves during the flowering stage.
ACTIONS: Urinary antiseptic.
USE: For any urinary infections, cystitis, urethritis. Particularly useful in situations where staling is painful. Use in combination with herbs such as marshmallow, couch grass, and yarrow for cystitis.
DOSE: 3-4 fresh or dried leaves daily.

BUCKWHEAT *Fagopyrum esculentum*

HABITAT: Native to Central Asia, but cultivated and naturalised in Europe, including Britain, and North America.
PARTS USED: Dried aerial parts for herbal applications, and seeds/nuts for flour production.
COLLECTION: During the flowering stage.
ACTIONS: Vasodilatory, for strengthening and repairing capillaries, antihistamine.
USE: Arthritis, poor circulation, haemorrhage of capillaries such as in epistaxis, navicular syndrome, any condition which could be helped by improving the blood supply, capillary fragility as a result of extensive X-rays.

Buckwheat is probably better known in the US as the source of buckwheat flour, used to make the famous buckwheat pancakes. In its herbal form it is excellent for any circulatory problems. Scientists at the Eastern Regional Research Laboratory of the US Department of Agriculture confirmed after extensive trials that rutin, which is one of the constituents of buckwheat, causes blood vessels to become more flexible and will strengthen fragile blood vessels. Back in the 1940s, US servicemen who had come into contact with atomic fission products or who had suffered from the effects of atomic radiation, were given rutin by the US Government.[6]

Buckwheat is rich in calcium, potassium, magnesium, iron and trace elements.

DOSE: 20-30 grams of the dried herb daily.

BURDOCK *Arctium lappa*

COMMON NAMES: Bardane, Beggar's Button.
HABITAT: Britain, Europe and North America. Found in fields and waysides, waste areas and around field borders.
PARTS USED: Roots.
COLLECTION: The roots should be unearthed in autumn.
ACTIONS: Bitters, alterative, diuretic, antiseptic.
USE: Arthritis, rheumatism, eczema. Being a bitters, burdock is an excellent digestive aid as it will stimulate the digestive juices. It is ideal for any blood disorders, liver and kidney function or toxic conditions which result in skin conditions such as eczema, sores, boils or dry scurfy skin. Burdock has been found to have anti-tumour activity and the root can be used as a poultice to speed up the healing of wounds. In the late 1800s the famous medical botanist Dr Withering recommended making a decoction of the plant for aching limbs!
DOSE: 10 grams daily.

CALENDULA *Calendula officinalis*

> 'The floures and leaves of Marigold being distilled...
> ceaseth the inflammation and taketh away the paine'
> *John Gerard 1597*

COMMON NAMES: Marigold, Pot Marigold, Marybud, Gold-bloom.
HABITAT: Native to Egypt and the Mediterranean and a common garden plant throughout the world.
PARTS USED: Petals and flower heads.
COLLECTION: Collect either the whole flower heads or just the petals, in

summer and autumn. Dry carefully to ensure the colour is retained.
ACTIONS: Astringent, anti-inflammatory, vulnerary, antiseptic, antifungal, cholagogue, emmenagogue.
USE: Traditionally used for horses as a blood tonic, and for stressed or fretful animals. Calendula is now more widely known and used for skin and gastric complaints. It contains essential oil and pro-vitamin A. Calendula tincture or extract is used in ointments and creams for cuts, bruises, burns and ulcers. It is used internally for gastric ulcers and inflammation, and being rich in sulphur has excellent blood-cleansing and antifungal actions. A lotion can be made with the flowers for bathing sore and inflamed eyes. Combined with clivers in equal proportions it is an excellent herb for the lymphatic and urinary system. It has been used internally in combination with clivers and nettle to treat skin conditions and urinary infections such as cystitis, which have failed to respond to conventional treatment.

There are a number of proprietary creams and ointments containing calendula available for general use on cuts, bruises and wounds.
DOSE: 15-20 grams of flowers daily in food, or brew 3 handfuls of flower heads with 1/2 litre/1 pint of boiling water and use as a lotion or compress for skin conditions.

CELERY SEED *Apium graveolens*

HABITAT: Widely cultivated; wild celery grows in marshy areas.
PARTS USED: Seeds and stems.
COLLECTION: In autumn, when the seeds are ripe.
ACTIONS: Antirheumatic, anti-inflammatory, carminative, diuretic, bitters, tonic.
USE: Arthritis, joint stiffness, urinary antiseptic. Horses can be given the celery stems, which will act as a tonic and diuretic. Celery seed is particularly good when used as a digestive tonic; it has a warming effect which will help if the horse is chilled or run-down, and its appetite reduced as a result (*see also* Fenugreek).

Celery is said to help with high blood pressure and to have a calming effect. The volatile oil, apiol, has an antifungal activity.[7]
DOSE: Dried seeds – approx 5-10 grams daily; stems – 2-3 daily.

CHAMOMILE, GERMAN *Matricaria recutita*

COMMON NAMES: Single Chamomile, Pin Heads.

HABITAT: Native to Europe, naturalised in N. America and extensively cultivated.

PARTS USED: Flowers.

COLLECTION: Collect the flowers throughout the summer. Do not pick them if they are wet, and take care to dry them carefully at not too high a temperature.

ACTIONS: Sedative, carminative, anti-inflammatory, relaxant, bitters, vasodilatory, analgesic, antispasmodic.

USE: For all cases of tension, restlessness and stress. Chamomile has been shown to have significant anti-inflammatory and analgesic actions, so is useful for aches and pains. A soothing lotion can be produced by adding a handful of the flowers to 1 litre/2 pints of boiling water and brewing. This liquid can then be used for bathing horse's eyes that have become sore or inflamed, and as a fomentation for mastitis. Chamomile tea is well known in human application for people who have difficulty sleeping.

DOSE: 1 handful of dried flowers in feed daily.

CLIVERS *Galium aparine*

COMMON NAMES: Cleavers, Goose-grass, Clives. Its Old English name was Catch-weed or Scratch-weed.

HABITAT: A rampant plant that is found in field and hedgerows throughout Europe.

PARTS USED: Leaf and stem, fresh and dried. It is distinguished by its hairy leaves and barbed stems which will cling on to anything in its reach.

COLLECTION: Must be collected before the plant flowers and seeds.

ACTIONS: Diuretic, astringent, aperient, alterative, tonic.

USE: Clivers is an excellent herb for supporting and toning the lymphatic system; it is specific for urinary infections such as cystitis, and enlarged lymph glands. It is rich in silica and as such will help strengthen coat and hair. It can be used for any soft swelling and fluid retention; being mildly diuretic it is particularly good for windgalls or filled legs.

Clivers is a much maligned plant, and one that is hated by farmers because of its rampant and choking nature. However, it has a great deal to offer the horse owner. If given the opportunity, horses will eat large quantities of this herb straight from the hedgerow. The fresh form can be given to horses with filled legs, and it is particularly good for laminitic ponies, as it can be cut and thrown into their 'lean' paddocks for them to chew on. Mixed in equal quantities with calendula it is the best

combination of herbs for supporting the lymphatic system.

I generally find that initially the horse is very enthusiastic about eating the herb, and then slowly loses interest in it over a period of a few days. This is normal and just the horse's way of telling you that he has had enough.

It is interesting that the homoeopathic remedy 'silica' is used to encourage elimination of any 'foreign bodies' that may have worked their way into the horse's system, such as grit, dirt, thorns, etc. By co-incidence silica-rich clivers have been found to be excellent in these cases as well.

DOSE: 2-3 handfuls of the fresh herb daily, or 20-30 grams of dried herb daily.

COMFREY *Symphytum officinale*

For 'inward wounds or burstings' and to
'cleanse the brest of phlegme and cureth the greefes of the lungs'
John Gerard

COMMON NAMES: Knitbone, Nipbone, Knitback, Brusewort.
HABITAT: Common throughout the UK, Europe and the USA. Found in moist areas and ditches. It has large hairy/prickly leaves and clumps of bell-like flowers. The colour can be as diverse as white-cream or blue-mauve-pink.
PARTS USED: Root and leaves, fresh or dried.
COLLECTION: Roots should be unearthed during spring or autumn. Leaves can be fed fresh or dried.
ACTIONS: Demulcent, anti-inflammatory, vulnerary, pulmonary, expectorant.
USE: Comfrey is one of the most widely used and famous of the healing plants. It is still used extensively as a fodder plant for horses and cattle in eastern Europe and Russia. In the past, as its common names suggest, it has been used for its remarkable ability to heal bone, cartilage and soft connective tissue. This is due to the presence of allantoin, which stimulates cell production, so encouraging wound healing, both internally and externally. Comfrey is an excellent source of vitamin B12.

Comfrey has been found to break down red blood cells, therefore supporting its use for bruising.

Research has shown that comfrey will reduce inflammation of the stomach lining, thereby making it ideal for any form of gastric disorder such as ulcers, colic and colitis.

Its pulmonary action is excellent for respiratory conditions, where it will soothe and reduce irritation as well as act as an expectorant.

It has been used traditionally as a remedy for rheumatism and arthritis. In Britain, the majority of old farm houses and cottages have it growing nearby.

DOSE: 1 handful of bruised fresh leaves daily (the reason for bruising the leaves is that they are very bristly/hairy and some horses will not eat them until they have wilted slightly), or 20-30 grams of dried leaf.

> **CAUTION**
>
> **It is advisable to avoid feeding the roots long-term – see page 34.**

EXTERNAL APPLICATIONS: Comfrey poultices, which are now available commercially, can be used externally as a poultice for ulcers and soft swellings. Fresh leaves can be bruised and put directly onto the swelling and then held gently in place with cool wet bandages. To make your own poultice, boil a good handful of chopped comfrey leaves wrapped in a piece of cotton or towelling, allow to cool, wring out the excess water (this water can be added to the horse's feed, if you wish) then apply to required area. For horses known to have a sensitive skin, it is worth noting that the poultice should be left on for no more than approximately 8-10 hours, and it is preferable not to re-poultice for another 8 hours at least.

I have used both comfrey oil and ointment (made by infusing the root and leaf in sunflower oil) with exceptional results. The ointment speeds up the healing process in wounds and reduces the risk of scar tissue; it is excellent for those stubborn wounds which refuse to heal. A word of caution, though: care should be taken when dealing with very deep external wounds as the comfrey is so effective it can lead to tissue forming on the top of the wound before the interior has healed.

I have used comfrey oil as a general muscle rub on tired and aching muscles; it has also proved very effective on hard bony growths – I massage these areas twice daily for 5 minutes each session. One of the old remedies used by racing yards was to massage splints for 10 minutes a day with either comfrey, olive or castor oil.

The Comfrey Controversy

I cannot pass on from comfrey without touching on the comfrey controversy which rears its ugly head every few years. There has been much confusion and disinformation about comfrey since one of the most damaging accounts concerning its use was published in 1968.[8] Depending on the bias of the writer, comfrey has been hailed as a 'cure-all' by some and 'dangerously toxic' by others. The truth would appear to be somewhere between these two extremes. The main concern is over the presence of pyrrolizidine alkaloids, or PAs for short. These alkaloids are present in ragwort and the borage family of plants, of which comfrey is a member. The PAs are present in higher concentrations in the root of the plant.

There are three main varieties of comfrey:
- **Common Comfrey** – *Symphytum officinale* – this is indigenous to the UK, Europe and the USA and the one most used in herbal preparations.
- **Prickly Comfrey** – *Symphytum asperum*.
- **Russian Comfrey** – *Symphytum x uplandicum* – Despite its name this variety originated in Sweden. It is a hybrid comfrey that was championed by organic gardening pioneer Henry Doubleday, who spent much of his life researching the herb for its ability to provide animal fodder and produce the ultimate compost and plant food (speak to any 'green' gardener!).

The concern rests on a number of reports based on experiments carried out on laboratory rats in Australia and Japan. It involved injecting rats with the alkaloid extracts from *Symphytum asperum* (not the common comfrey which is the UK's indigenous species and the variety most commonly used in herbal preparations.) The purified symphytine (which is the carcinogenic alkaloid found in comfrey) was injected into rats at a rate of 300mg per kg bodyweight and represented 33% of their food intake.[8] This, apart from anything else, would be sufficient to lead to protein deficiency and result in malnutrition in the rats. Fifteen per cent of the rats developed liver cancer and 50% died at a point when the comfrey intake was equivalent to a human eating 5,607 leaves![9] To reproduce this experiment in horses, an average-size horse weighing 500kg/1100lbs would need to ingest 150kg/330lbs of the pure symphytine alkaloid, which represents 5% of the total alkaloids present in comfrey. I will leave you to do the arithmetic.

This information puts the scare stories into some kind of perspective. The UK's Ministry of Agriculture has recently made a recommendation that comfrey root should not be used in feed products for horses, and that comfrey leaf should not be given to horses on a long-term basis. My own feeling is that, like most things, if given in moderation as and when necessary, comfrey presents no greater threat than any other herb, and for external use there is nothing to touch it.

COUCH GRASS *Agropyron repens*

COMMON NAMES: Twitch Grass, Scutch, Witchgrass, Quick Grass, Dog's Grass (USA).
HABITAT: Grows all over the world, including Britain and Europe.
PARTS USED: Leaves and rhizomes.
COLLECTION: Unearth the rhizomes either in spring or autumn, dry and chop for use. Leaves, anytime.
ACTIONS: Diuretic, demulcent, aperient, antibiotic.
USE: Horses are just one of the animals that are especially fond of couch grass and will seek it out for use as a spring tonic. It is excellent for use as a urinary antiseptic (*see also* Uva-Ursi), particularly for cystitis and kidney infections, and for constipation. The plant is rich in silica, which will help strengthen coat and hooves, as well as iron, mucilage and vitamins A and B. The plant contains a volatile oil which has been shown to have antibiotic properties.
DOSE: Let horses graze at will on the fresh grass; if this is not possible give several handfuls of the leaves daily. Either fresh or dried cut roots are best used for urinary problems – give 20-30 grams per day.

DANDELION *Taraxacum officinale*

COMMON NAMES: Fairy Clock, Pee-the-Bed, Lion's Teeth.
HABITAT: Distributed throughout the majority of the world.
PARTS USED: Root and leaves.
COLLECTION: Leaves anytime; roots should be collected in mid to late summer.
ACTIONS: Diuretic, mild laxative, tonic, antirheumatic, hepatic.
USE: For any liver or kidney disorder, including jaundice. The dandelion root has the greater hepatic action, being used to stimulate the liver and increase the production of bile. The leaves have a powerful diuretic action, hence its Old English name – Pee-the-Bed. The added benefit that dandelion has when used as a diuretic is that it is rich in potassium, magnesium and calcium. The rich potassium content replaces the potassium that is leached from the body. Dandelion is rich in vitamins A, B, C and D, and contains higher levels of vitamin A than carrots.[10] Dandelion will cleanse the blood and is useful in the treatment of rheumatism.

The leaves can be used in salads, the flowers make a lovely wine and the white milky sap is supposed to be a cure for warts for people and horses. I look upon the dandelion as a natural electrolyte and have had many reports of horses digging up their paddocks to get to the roots.

DOSE: Several handfuls of fresh leaves daily; 30 grams dried leaf daily; 4-5 fresh roots daily; 20 grams dried root daily.

DEVIL'S CLAW *Harpogophytum procumbens*

HABITAT: Particularly found in Namibia, but indigenous to South and Eastern Africa.

PARTS USED: Root.

COLLECTION: The roots are collected after the rainy season.

ACTIONS: Anti-inflammatory, analgesic, sedative, antirheumatic, digestive stimulant.

USE: In cases of arthritis when there is pain or inflammation, any form of degenerative joint disorder, or bony changes resulting in inflammation. Tests carried out in Germany (one of the biggest users of devil's claw) have shown the analgesic and anti-inflammatory effects to be comparable with cortisone and phenylbutazone, without the attendant side-effects.[11] Recently a pure herbal product containing devil's claw, willow and other herbs has been trialled by a number of veterinary practices in the UK as an alternative to phenylbutazone. Further tests suggest that the plant is also diuretic and stimulant to the liver and because of its bitters content it can help to encourage appetite. Avoid using the herb when gastric ulcers are indicated.

DOSE: 15 grams of cut root daily.

> **CAUTION**
>
> **DO NOT GIVE TO MARES IN FOAL.** In human herbal medicine, devil's claw is said to be a uterine muscle stimulant, so for safety's sake it is advisable to avoid using the herb during pregnancy.

ECHINACEA *Echinacea angustifolia, E. purpurea*

COMMON NAMES: American Purple Coneflower, Black Sampson, Missouri Snakeroot, Kansas Snakeroot.

HABITAT: Indigenous to northern USA, cultivated throughout Europe.

PARTS USED: Roots.

COLLECTION: The tops of *Echinacea purpurea* can be used throughout the summer. The roots of both species should be collected in the autumn, just before the first frosts.

ACTIONS: Antiviral, antibacterial, immuno-stimulant, anti-inflammatory, vulnerary.

USE: For chronic viral and bacterial infections, and depressed immune systems. For skin complaints and to encourage wound healing in general. Echinacea can be used internally, and externally as a poultice or compress. As a prophylactic, this herb can be used to help protect horses from infections, and is ideal when used in this way for yards where there is a risk of an outbreak of strangles or viral infection. Use also for urinary infections such as cystitis and urethritis.[12]

This is truly a wonderful herb which has recently been used with great success on horses who appear to have the symptoms of post-viral syndrome. The owner of a showjumper which had gone down with a mystery virus, was at her wits end, when two years after the virus had struck, the horse was still not fully recovered. Despite complete rest, good food and care, whenever the horse was asked to do any sort of exercise, however gentle, he would show signs of lethargy, depression and a disinclination to work; in addition, blood tests revealed a high white cell count, indicating that the horse's system was still not right. The horse was started on a programme of echinacea, feeding only 10 grams a day. Within a few weeks the owner reported a slight improvement in the horse, which

was confirmed by a blood test. Over the next six months the horse, whilst continuing to receive his daily dose of echinacea, was reintroduced to work. During this time regular blood tests were taken which confirmed that his blood profile had stabilised. At the end of this time the owner discontinued the echinacea, the horse remained stable, and was returned to full work.

Echinacea was first used by the native American Indian tribes (Indians would suck on a piece of the root all day) and was first classified in Europe in the 1690s. However, the herb did not appear in a medical journal until 1891. Since this time it has been the subject of over four hundred scientific articles. It has been found to be an effective immuno-stimulant, in that it stimulates the white blood cells to help fight infection, and increases the power of the immune cells to better help the body fight bacteria and infected cells. It is no coincidence that the 'AIDS' programme in the US is investigating the use of echinacea.

I am not able to cover in depth the range of uses of this wonderful herb, so for readers interested in knowing more about it and its applications some further reading is advisable. (*See* Bibliography and Further Reading section.)

DOSE: 10-20 grams of cut root daily.

EYEBRIGHT *Euphrasia officinalis*

HABITAT: Meadowland throughout Britain and Europe.

PARTS USED: Aerial parts.

COLLECTION: Gather the whole plant whilst in bloom during late summer or autumn.

ACTIONS: Astringent, anti-catarrhal, anti-inflammatory.

USE: Specific for any problems of mucous membranes. Use internally for persistent catarrh, particularly nasal. It is best known for its use on any eye conditions such as inflammation, weeping, stinging, ulceration, conjunctivitis. Eyebright has been used very successfully, both internally and externally, on both dogs and horses that suffer from persistent ulceration of the eye. It can be made into a tea and used as an eye-wash or compress, and is excellent when combined with distilled witch hazel.

DOSE: For an eye-wash brew 30 grams of the herb in $^1/_2$ litre/1 pint of boiling water. Use to bathe eyes and soak compresses. Internally give 20-30 grams of the dried herb daily.

FENUGREEK *Trigonella foenum-graecum*

COMMON NAMES: Foenugreek, Bird's Foot, Greek Hay-seed.
HABITAT: North Africa and India, and cultivated worldwide, including small quantities in the UK.
PARTS USED: Usually the seed, although the aerial parts are used as a fodder crop for animals in many parts of the world.
COLLECTION: The seeds are harvested in the autumn.
ACTIONS: Demulcent, nutritive, laxative, galactogogue, expectorant.
USE: The name comes from the Latin *foenum-graecum*, meaning 'Greek hay' - this is due to the fact that in the past the plant was used to add 'nose' to inferior or poor hay. In Mrs M. Grieve's *A Modern Herbal*, written in 1931, she makes reference to the fact that fenugreek was used extensively at that time as a feed additive for horses and cattle, to help put on condition. It is interesting to note that fenugreek's chemical composition is very similar to that of cod-liver oil.

Fenugreek can be used to stimulate appetite, for gastric disorders, ulceration, poor milk production, and general condition. Probably better known as a culinary herb used for flavouring curries, fenugreek is a trifoliate like alfalfa and clover. It contains what are known as steroidal saponins, which have proven to be anti-diabetic in animals, and, like its 'sister' herb garlic, it reduces cholesterol levels in blood. These saponins closely resemble the body's sex hormones, which accounts for its traditional use as an aphrodisiac and to help increase milk production. The seeds are extremely nutritive, being made up of 8% oil and 20% protein, and containing vitamins A, B and C, and the fertility vitamin E. They are rich in calcium, which is vital during nursing. Because of its pleasant aroma, fenugreek is ideal to help tempt shy feeders, and to put condition on 'poor doers'. This herb is particularly good when combined with garlic, as the two herbs complement each other, helping to release all their potency.
DOSE: 20-30 grams of the seed daily.

> ### NOTE
> **I tend to avoid using fenugreek on mares that may have an hormonal imbalance** – when given in large quantities, there is a possibility of uncharacteristic behaviour, possibly due to the herb's oestrogen content.

GARLIC *Allium sativum*

HABITAT: Cultivated throughout the world.

COLLECTION: Collect the bulbs when the leaves begin to yellow and droop.

ACTIONS: Antiseptic, antibiotic, antimicrobial, anthelmintic, expectorant, diaphoretic, hypotensive, anti-diabetic, anti-thrombotic.

USE: 'One of the Plague herbs' and the best known and most widely used herb in the horse world. Garlic is now included in many of the proprietary brands of horse coarse mixes, as well as being available in powdered form for addition to feed. Garlic is ideal for any respiratory disorder; being both expectorant and antibiotic it will encourage the expulsion of mucus from the lungs and help with any infection present. Rich in sulphur, which is excreted through the pores of the skin, garlic cleanses the blood and when excreted will help to deter biting flies. With regard to digestion, garlic supports the development of natural bacterial flora, so vital to a good digestive system, whilst killing pathogens. Garlic is one of the best herbs to use when wishing to protect animals or humans from infection. Used in a prophylactic way internally it can guard against coughs, viral infection, and worm infestation. Externally the juice from a cut bulb can be used on cuts, bites, stings, ringworm, lice, and tick bites.

NB: Care should be taken if you are feeding garlic to nursing mares as it may taint and 'come through' the milk. This may upset the delicate digestive system of young foals.

Mrs M. Grieve's *A Modern Herbal*, published in the 1930s, contains the following interesting anecdote:

> 'There is a curious superstition in some parts of Europe, that if a morsel of the bulb be chewed by a man running a race it will prevent his competitors from getting ahead of him, and Hungarian jockeys will sometimes fasten a clove of garlic to the bits of their horses in the belief that any other racers running close to those thus baited, will fall back the moment they smell the offensive odour!'[13]

(*See* Bibliography and Further Reading)

DOSE: 6-8 fresh, crushed cloves daily, or 15-30 grams of pure garlic powder daily.

GOLDEN ROD *Solidago virgaurea*

COMMON NAMES: Woundwort, Aaron's Rod. The name Solidago comes from the Latin *solidare,* which means to unite. In the past its healing powers were so highly regarded that large quantities of the plant were brought in from the Continent to Britain for medicinal use, before botanists discovered it growing in Britain.
HABITAT: A common garden plant throughout Britain, Europe and North America.
PARTS USED: Leaves and flowering tops.
COLLECTION: Early summer through to autumn, or when just coming into flower.
ACTIONS: Anti-inflammatory, antiseptic, diuretic, diaphoretic, carminative, antifungal, astringent.
USE: For urinary infections, and kidney stones. Golden rod is an excellent digestive aid and is useful for bathing wounds, especially if infected. Use in combination with echinacea for upper respiratory tract infections and catarrh.
DOSE: 30 grams daily of the dried herb, or 1-2 handfuls of the fresh herb. Golden rod makes a very pleasant tea, and can be used in feed to help with digestion. Add 1 tablespoon of herb to one cup of boiling water and allow to brew.

HAWTHORN *Crataegus oxyacantha*

> 'Mark the faire blooming of the hawthorne tree,
> Who, finely cloathed in a robe of white,
> Fills full the wanton eye with May's delight'
> *Chaucer*

COMMON NAMES: May, Haw, Whitethorn, Hagthorn.
HABITAT: Common in Britain, Europe and USA.
PARTS USED: Berries, leaf tops and flowers.
COLLECTION: Spring – flowers and leaves; summer – leaves and berries; autumn – berries.
ACTIONS: Cardiac tonic, vasodilatory, hypotensive, antisclerotic.
USE: The hawthorn is well known in the UK and is steeped in folklore. It was said that a mare's afterbirth should be draped over either a hawthorn or a holly to ensure good luck.

Hawthorn leaves, flowers and berries, have a beneficial action on blood flow, blood pressure and heart rate, and hawthorn is one of the best tonics for the heart and circulatory system. The shrub contains what are known

as flavonoids (nineteen in all have been found, including rutin – *see* Buckwheat). These flavonoids have the distinct action of dilating the peripheral blood vessels (an action very often induced with drugs in cases of navicular syndrome), which will improve blood supply to affected areas. The incredible feature of this plant is that it has the ability to normalise blood pressure. It is used in human herbal medicine for treating angina, irregular heartbeat and arteriosclerosis. I have known horses and ponies that were suffering from conditions such as navicular and laminitis, to repeatedly seek out hawthorn and pick off the sprouting leaf tops whenever they could.

In the USA and Britain the berries and leaves are used, whilst in Europe the leaves or flowers are preferred.

DOSE: 1 handful of fresh leaves or sprouting buds daily, or 10-12 grams of berries daily.

HOREHOUND *Marrubium vulgare*

COMMON NAMES: White Horehound, Hoarhound.
HABITAT: Grows wild on wasteland and roadsides throughout central and southern Europe, scattered throughout Britain and Northern America.
PARTS USED: Dried leaves and flower tops.
COLLECTION: Collect the aerial parts in mid to late summer.
ACTIONS: Expectorant, circulatory stimulant, bitter digestive and hepatic tonic.
USE: Horehound has traditionally been used in tonics, cough candies and ale. Its expectorant action makes it a specific for respiratory conditions, particularly for treatment of chesty non-productive coughs, and catarrh of the respiratory tract. It is a gentle digestive stimulant and can be used to encourage a poor appetite; however, some horses dislike its bitter taste.

DOSE: 1-2 handfuls of the fresh herb daily, either in feed or made into a brew with 1/2-1 litre/1-2 pints of water, or 15-20 grams of the dried herb daily in feed.

HORSERADISH *Amoracia rusticana*

HABITAT: Grows wild in Eastern Europe; grows wild and is also cultivated in Britain and the US.
PARTS USED: Root and leaves.
COLLECTION: Collect roots in the autumn and store in sand. Leaves, anytime.
ACTIONS: Anthelmintic, stimulant, rubefacient, diuretic, mild laxative.
USE: Horseradish used internally and externally is excellent for poor circulation. It is a powerful expellant of worms and can be used in cases of urinary infection. It is an old remedy for fevers and influenza. It will encourage a poor appetite and is an excellent stimulant.

I have used horseradish very successfully on my own horses to reduce stubborn windgalls. I have also found it very useful in conjunction with ginger and turmeric as a poultice to help disperse both hard and soft swellings. For details, see page 21.

I have horseradish growing in my garden and my horse loves to eat the occasional fresh picked leaf.
DOSE: One small fresh root as needed, grated and mixed into a bran mash, or 5 grams of powdered root as needed. Leaves, occasionally.

HYPERICUM *Hypericum perforatum*

COMMON NAME: St John's Wort.
HABITAT: Throughout Britain, Europe, North America and Asia. It grows in open woodland, hedgerows and banks.
PARTS USED: Aerial parts.
COLLECTION: Collect the whole stem when in flower and dry carefully.
ACTIONS: Anti-inflammatory, sedative, astringent, vulnerary.
USE: Used externally, the herb is excellent for wounds. The oil which is

made from the flowers has analgesic and antiseptic properties and can be used for rheumatic pain, burns and skin conditions. Do not use the neat oil on broken skin. A few drops of hypericum tincture added to warm water is ideal for washing and cleansing wounds.

DOSE: Hypericum is not recommended for internal use.

CAUTION

Horses who are known to suffer from photosensitivity may react to hypericum products if used internally. I did hear from a woman whose pink-skinned cow had eaten hypericum in the field and suffered an extreme reaction.

KELP *Fucus vesiculosus*

COMMON NAMES: Bladderwrack, Kelpware, Rockweed, Seawrack.
HABITAT: Common in cold seas, including the North Sea and Atlantic Ocean.
PARTS USED: Whole plant (usually dried in granulated form).
COLLECTION: From any area where the sea is known to be pollution-free (if such as place exists these days). Kelp has been found to accumulate toxic waste and heavy metals such as cadmium and strontium when grown in polluted areas. It is therefore vital that you ascertain the source of the kelp before using it.
ACTIONS: Antihypothyroid, antirheumatic, antibiotic.
USE: Kelp is the original source of iodine. Used extensively for underactive thyroid, and as an excellent source of minerals for the body. Can be used externally for compresses to reduce inflammation and arthritic pain. Remember: kelp has a strong flavour which should be introduced gradually to the horse's feed.
DOSE: 15-30 grams daily in feed.

LAVENDER *Lavendula angustifolia*

COMMON NAMES: Garden Lavender, Spike Lavender.
HABITAT: Native to Southern Europe, cultivated widely, especially in the Provence region of southern France, and in the Channel Islands. English lavender has the reputation of being the best scented.
PARTS USED: Flowers.
COLLECTION: Throughout the summer, just before the flowers open.
ACTIONS: Relaxant, antispasmodic, antidepressant, circulatory stimulant, antimicrobial.
USE: Lavender is used in essential oil form and is for external use only. (For further details see Aromatherapy section.)
NB: See also competition warning on page 146.
DOSE: Lavender is not recommended for internal use.

LIQUORICE *Glycyrrhiza glabra*

COMMON NAMES: Licorice, Spanish or Italian Liquorice.
HABITAT: Native to south-east Europe and parts of Asia. Cultivated extensively in Britain up until the 1940s when the land was needed for the war effort, liquorice was used to make sweets and cough syrups. This became a large industry in northern England and many people have childhood memories of the famous liquorice sweets, Pontefract Cakes, named after the Yorkshire town which made them.
PARTS USED: Dried root, occasionally the leaves.
COLLECTION: Unearth the roots in late autumn and dry.
ACTIONS: Expectorant, demulcent, anti-inflammatory, laxative, spasmolytic, antitussive, antibacterial, antiviral.
USE: Liquorice has been used medicinally for over 3000 years, and was spoken of by the ancient Greeks in the third century BC; ancient herbals carry instructions for its use on both animals and humans. The native American Indians used liquorice leaves on their horses to help heal sores; however, it is the root that is most frequently used. Liquorice has an oestrogenic activity and has been used to improve fertility in female animals; it is therefore advisable to avoid its use during pregnancy. The main active ingredient is glycyrrhizin, which is responsible for the sweet taste; it is a non-sucrose sweetner and is fifty times sweeter than sucrose!

Liquorice's demulcent action is excellent for reducing gastric acid secretions. This makes it a specific for gastric ulceration and inflammation. For respiratory conditions, liquorice has both a demulcent and expectorant action by loosening tracheal mucous secretions. It is anti-inflammatory and has shown an ability to improve liver function. In

Chinese medicine it is used extensively for liver disorders such as hepatitis and jaundice.
DOSE: 10-20 grams daily.

> **CAUTION**
>
> **Do not overfeed.** Although not proven, it has been suggested that in excessive quantities liquorice can lead to potassium depletion, fluid retention and weight gain.

MARSHMALLOW *Althea officinalis*

COMMON NAMES: Mallards, Mauls.
HABITAT: Originally in uncultivated marshy ground in southern Europe and Asia, now throughout Europe, North America and Australia.
PARTS USED: Root and leaves.
COLLECTION: Collect the leaves in summer after the plant has flowered. Collect the roots in late autumn when the mucilage content is at its highest.
ACTIONS: Root – demulcent, emollient, diuretic; leaves – expectorant, demulcent, emollient and diuretic.
USE: Both the leaves and root can be used for coughs, due to their expectorant, emollient and mucilagenous qualities. They will encourage the expulsion of mucus, particularly when the cough is dry and tight, and soothe irritation in the airways. An excellent herb for digestive and urinary conditions. The root is particularly high in mucilage and is ideal for use where there is irritation or inflammation. It is specific for any condition of the digestive system such as gastric ulceration, inflammation and colitis, as well as inflammation of the urinary passages in conditions such as cystitis. Externally the root can be used as a poultice for drawing infection and soothing skin conditions (mix equal parts of powdered marshmallow root and slippery elm powder). Powdered marshmallow root has been used as a preventative on horses prone to spasmodic colic, and on horses and dogs for cystitis and kidney stones.
DOSE: 1 x 15ml scoop of powdered marshmallow daily, or 20 grams of cut root daily, in feed.

MEADOWSWEET *Filipendula ulmaria*

COMMON NAMES: Queen of the Meadow, Bridewort, Lady of the Meadow.
HABITAT: Common throughout Britain and Europe, in parts of Asia and Northern America.
PARTS USED: Aerial parts.
COLLECTION: Collect the fully opened flowers and leaves throughout the summer. The small white flowers have a powerful almond scent.
ACTIONS: Astringent, antirheumatic, antacid, stomachic.
USE: A specific for gastric ulcers, meadowsweet is particularly efficacious against ulceration caused by drugs. It is one of the best herbs for the digestive system; it will protect the digestive tract and reduce excess acid. Its gentle astringent action will help with scouring. Also ideal for fevers and rheumatic pain.

Meadowsweet is the herbal aspirin – but better, and without any of the side-effects! Meadowsweet contains a substance called salicylic acid which is found in the flower buds; the same substance is also found in the bark of the willow – *Salix alba*. In the late 1890s the pharmaceuticals company Bayer formulated a new drug called acetylsalicylic acid, which we know better as aspirin. In fact, the name aspirin is said to have been created from the letters of the Old English name for meadowsweet – *Spiraea ulmaria*.

It is the salicylates in meadowsweet that have the anti-inflammatory action on rheumatic pain and fever, as well as being antiseptic and diuretic. These salicylates are balanced by the other constituents in the plant, such as the tannins and mucilage. The salicylates in isolation can cause gastric bleeding – a now well-known possible side-effect of aspirin. However when the plant is left as a whole, in balance with its other constituents, then you have the opposite effect – a herb which is actually used to heal gastric bleeding and ulceration! Thus meadowsweet is an excellent example of the whole being better than its isolated components, and an example of how, when man interferes with nature he creates imbalance.
DOSE: Two handfuls of fresh herb, cut small and mixed in feed or made into a brew with 1 litre/2 pints of boiling water; or 20-30 grams of dried herb daily, in feed.

MILK THISTLE *Silybum marianum*

COMMON NAMES: Marian Thistle, Mary's Thistle.
HABITAT: Originated in western and central Europe, but now grows wild in dry and rocky places in southern and western Europe, throughout the USA, South America and Australia. It is easily cultivated.

PARTS USED: Seeds.

COLLECTION: The thistle heads should be collected (remember to wear thick protective gloves!) when they have finished flowering and are showing the white 'parachutes'. This shows that the head contains ripe seeds. The ripe seeds have the highest content of silymarin.

ACTIONS: Hepatoprotective, demulcent, cholagogue.

USE: As its name suggests, milk thistle has been used historically to promote milk production, although I have no experience of its use on horses in this context. Milk thistle is most famous for its ability to protect and improve liver function, and to speed up the regeneration of liver cells. It increases bile secretions from the liver. Milk thistle seed has undergone extensive trials in Europe and the US. It has been shown to lower fat deposits in the liver of animals and to be able to protect the liver from damage by drugs and poisons. Milk thistle seed has been used on horses and ponies that are suffering from liver damage due to worm burden, or prolonged use of drugs. It is an excellent herb to be given in the spring as a tonic and to help strengthen the liver. It is slowly absorbed by the body, so it is necessary to give the seeds for 4-6 weeks. It is interesting to note that the seeds of the milk thistle are high in linoleic acid – up to 60%. It is this acid that has proved so successful in helping to regulate female hormonal balance and is better known as being the active ingredient in evening primrose oil.

Many horse owners will have noticed their horses displaying cravings for the common thistle, a close relative of the milk thistle, particularly in spring and winter. The perseverance they display in reaching and eating this plant, despite the obvious difficulties, is clear proof that the horse is acting on long-established instincts. Milk thistle is not common in Britain, and an old legend has it, that where it is found, it marks the site of a landing by shipwrecked sailors, who brought the seeds with them from places like Spain and Turkey.

DOSE: 10-15 grams of seeds daily. If the seeds can be ground in a coffee grinder or food processor before use, so much the better.

MINT *Mentha piperita*

COMMON NAMES: There are many varieties of mint who share the same active constituents. These include: Peppermint, Spearmint, Watermint, Apple Mint, and Black or Mitcham Mint (so called because the first known cultivation of peppermint was in Mitcham, near London, where in 1850 over 500 acres were under cultivation).

HABITAT: Britain, Europe, USA, Asia and N. Africa.

PARTS USED: Leaves.

COLLECTION: Collect the leaves throughout the growing season.
ACTIONS: Stimulant, antispasmodic, carminative, anti-inflammatory.
USE: Like garlic, mint is now used extensively as an additive in commercial horse feeds. This is because its delightful aroma makes the feed more appetising and it has a beneficial effect on the digestive system. It is one of the best digestive aids available and helps soothe and relax the digestive tract. The volatile oil it contains is a specific for ulcers. The flavonoids it contains will stimulate the liver, increasing the flow of bile. Menthol is one of the components of the volatile oil contained in mint and is antibacterial and antiparasitic. The oil is excellent for flatulence and colic. (Goats suffer badly from flatulence and many goat owners give their animals small quantities of peppermint oil to help with this problem.) Peppermint oil (3 or 4 drops) can also be sprinkled onto hay or bran in a nosebag and used as an inhalant to help relieve upper respiratory problems and catarrh. Mint can be used to help dry off milk in nursing mares.
DOSE: 15-20 grams of dried herb daily, or 1-2 handfuls of the fresh herb daily. The watermint variety grows freely in moist places and horses will eat this readily if given access to it.

NETTLE *Urtica dioica*

'Tender handed, grasp the nettle
And it stings you for your pains
Grasp it like a man of mettle
And it soft as silk remains'

COMMON NAME: Stinging Nettle.
HABITAT: Throughout the world on waste ground, hedgerows or anywhere where the earth is disturbed.
PARTS USED: Aerial parts.
COLLECTION: Any time, but preferably when the plant is in flower.
ACTIONS: Diuretic, astringent, haemostatic, circulatory stimulant.
USE: Nettles are a rich source of vitamin C, iron, sodium, chlorophyll, protein and dietary fibre. This makes them ideal for use as a spring tonic and general blood cleanser and conditioner. Excellent for anaemia, because of their high iron and vitamin C content, which ensures that the iron is absorbed efficiently by the body. Nettles will stimulate the circulation, which makes them ideal for conditions such as laminitis, rheumatism and arthritis.

A wash made from the nettle can be used as a rinse for horses' manes and tails; and when nettles are fed internally they encourage 'dappling' on a horse's coat. The easiest way to feed nettles when they are growing freely is to cut them, leave them to wilt in the sun to allow the 'sting' to go out of them, then cut them up and add to the horse's feed, or simply throw the wilted nettles into the horse's paddock to graze on. Remember: nettles are a blood tonic, and some horse owners may find that they have the action of livening up their horse!

DOSE: 20-30 grams of cut, dried herb daily.

> ### CAUTION
>
> **Some horses can react to nettles and will come out in a 'nettle rash'.** This is shown as small raised lumps under the horse's coat. Luckily this condition is not long-lasting and will normally disappear within 24 hours. However, if you know your horse is sensitive then be cautious, and discontinue use if a reaction occurs.

PARSLEY *Petroselinum crispum*

HABITAT: Indigenous to the eastern Mediterranean, but cultivated and naturalised throughout the world.

PARTS USED: Roots, leaves and seeds.

COLLECTION: Leaves can be gathered at any time during the growing season. The roots are collected in the autumn from plants that are at least two years old.

ACTIONS: Diuretic, carminative, uterine stimulant, expectorant, digestive tonic.

USE: Parsley is a strong diuretic as well as being rich in vitamin C, iron and copper, which makes it ideal as a supplement for anaemia. It will encourage milk production in nursing mares. It can be used for rheumatic conditions, as well as being a specific for urinary infections or stones.

Tradition has it that parsley was used by horsemen to calm difficult or over-excitable horses. However, Homer states that the ancient Greek charioteers fed parsley to their horses to pep them up! The constituents of the plant certainly would not suggest that parsley should have a calming action; however, I have spoken to several people who have used it successfully for this problem. One has to consider the possibility that the horse could have been behaving badly because it was suffering from a urinary or digestive problem and the active constituents of the parsley eased the discomfort, resulting in a calmer, happier horse. As I have said elsewhere, it is always best to try to ascertain the cause of the problem rather than just suppress the symptoms.

DOSE: Two or three handfuls of the fresh herb daily, or 2 fresh roots daily, or 20-30 grams of dried parsley daily.

> **NOTE**
>
> **As parsley has uterine stimulatory properties do not give it to mares in foal.**

POPPY *Papaver rhoeas*

COMMON NAMES: Red Poppy, Corn Poppy, Corn Rose.
HABITAT: Common throughout Europe and Britain, in fields and on disturbed ground.
PARTS USED: Seeds and flowers.
COLLECTION: Collect the petals when dry in summer/autumn. Collect the seeds from the dried seed head in autumn.
ACTIONS: Mild sedative, expectorant, anodyne.
USE: Do not confuse the corn poppy with its relative the opium poppy. The corn poppy does not have the potency of the opium poppy. It is, however, very effective in calming excitable and fretful horses. Poppy petals are excellent to help soothe inflamed throats and as an expectorant for persistent catarrh and irritable coughs. The seeds are a tonic for horses.
DOSE: 5 grams of dried petals daily; 5-10 grams of seed occasionally as a tonic.

RASPBERRY *Rubus idaeus*

HABITAT: Throughout Britain, Europe, USA and northern Asia, cultivated in temperate climates.
PARTS USED: Leaves.
COLLECTION: Collect the leaves throughout the growing season.
ACTIONS: Astringent, tonic, parturient.
USE: Traditionally used for animals (and humans) to help strengthen and tone the uterine muscles, to assist with contractions during foaling, checking haemorrhage and encouraging cleansing of the afterbirth. Some people prefer to give small quantities of raspberry leaves regularly throughout pregnancy, especially if there is a danger of haemorrhage.

However, it is usually given for the last 4-6 weeks of pregnancy, and after foaling for 1 week to help encourage cleansing and toning of the uterus.

Raspberry leaves can also be made into a brew for inflammation and ulceration of the mouth.

DOSE: 20-30 grams daily.

RED CLOVER *Trifolium pratense*

HABITAT: Throughout Britain and Europe, naturalised in N. America and widespread throughout the world.
PARTS USED: Flowers.
COLLECTION: Collect the flower heads throughout the flowering season, generally spring to late summer.
ACTIONS: Alterative, expectorant, diuretic, antispasmodic, oestrogenic.
USE: All varieties of clover have a sedative and calming action, and yet clover can be used as a tonic. It is especially beneficial for horses that are run down or recovering from viral illnesses, in particular if a cough or mucus is one of the symptoms. A brew of clover heads can be made for external use on skin conditions such as eczema. Red clover is said to have anti-tumour activity and has been used in the herbal treatment of cancers; however, very little research has been carried out to support this.
DOSE: 1-2 handfuls of the flower heads daily, with feed. Two handfuls of flowers with 1/2 litre/1 pint of boiling water.

> ### CAUTION
>
> **In large quantities red clover can have an oestrogenic activity in horses.** Very often, horses turned out onto pasture intended for cattle will find the high clover levels too much to cope with. I have had experience of horses developing a condition similar to mud fever as a result of eating clover, and other trifoliates. (*See* 'Important Note' in Mud Fever section, page 114.)

ROSEHIP *Rosa canina*

COMMON NAMES: Dog Rose, Wild Briar, Briar Rose.

HABITAT: Throughout Britain, Europe, N. Africa, parts of Asia, naturalised on the east coast of the US and cultivated extensively elsewhere in the world.

PARTS USED: The fruits (hips).

COLLECTION: Collect the hips in the autumn.

ACTIONS: Astringent, aperient, anti-diarrhoeal.

USE: *Rosa canina*, the dog-rose, is so named because it was believed that the root was a cure for rabies. Most of us will know it better as the source of rosehip syrup which children have grown up on since the 1940s. The hips are a rich source of vitamin C (approximately 3000mg per kg, although in a recent analysis a figure of 6000mg per kg was reported) as well as vitamin A, thiamine, niacin, riboflavin, vitamin K and volatile oil. Excellent for use as a spring tonic or in cases of general debility, rosehips are a mild astringent, and will help with scouring. They have been used most successfully as part of a mixture of herbs to help horses return to full health after lengthy illnesses. The high vitamin C levels will help the horse fight off infection and restore its defences. I also use rosehips to help encourage strong, healthy hoof growth. I originally believed this was due to the biotin content. However, since having them analysed and finding that the biotin content is negligible, I now feel that it is probably the flavonoids, along with the high vitamin levels, that must be responsible for this effect.

DOSE: 15-20 grams daily of chopped rosehip shells.

ROSEMARY *Rosmarinus officinalis*

HABITAT: Indigenous to southern Europe, easily and widely cultivated throughout the world.

PARTS USED: Leaves.

COLLECTION: The leaves are at their best during the flowering period; however, they can be collected at any time during the summer.

ACTIONS: Anti-inflammatory, tonic, antiseptic, stimulant, rubefacient, antispasmodic, carminative.

USE: To stimulate circulation, and as a blood cleanser. Rosemary is also excellent for toning and calming the digestive system, particularly when stress and nervous tension are the cause. A specific for rheumatism and liver function, its flavonoid content will help reduce capillary fragility.

Due to its essential oil content, rosemary has excellent antibacterial and antifungal properties (it was one of the herbs used during the Great Plague of London (1665) to help repel infections, and thirteenth-century herbalists used it to cure headaches, bad dreams and insanity!). It is still recommended as a cure for headaches and is used in shampoos and hair washes. Rosemary has a wonderful fragrance and the essential oil will calm and soothe.

DOSE: 1 handful of the leaves daily, or 15 grams of dried herb daily, in feed.

CAUTIONS

Do not give to mares in foal.

Also, many of the volatile oils found in fragrant herbs such as rosemary contain oils such as thymol and camphor. If the essential oil is used as part of an external application, care should be taken to discontinue its use 10 days prior to competing under Jockey Club or FEI rules. If the oil is licked and ingested by the horse, although not physically harmful, it may be detected in blood tests and be regarded as a prohibited substance.

SAGE *Salva officinalis*

COMMON NAMES: Red Sage, Garden Sage.
HABITAT: Indigenous to the Mediterranean region, but cultivated throughout the world.
PARTS USED: Leaves.
COLLECTION: Collect the leaves throughout the summer.
ACTIONS: Astringent, antiseptic, aromatic, uterine stimulant.
USE: Sage is particularly useful for any sort of infection or soreness of the mouth. It has been used effectively for mouth ulcers in horses, and also as one of the ingredients in a herbal 'antibiotic' mix, which has had great success in horses that have stopped responding to conventional antibiotics. The volatile oil it contains makes it particularly effective for digestive function and as a digestive carminative. It is also useful for reducing fevers in horses with 'flu. Sage can reduce the milk flow in nursing mares, if for some reason this is necessary.
DOSE: 5 grams daily of chopped leaves mixed in feed. If using for mouth infections or soreness, mix with honey.

> **CAUTION**
>
> **Do not give to mares in foal. Use in small quantities and not for prolonged periods.**

SCULLCAP *Scutellaria laterifolia*

COMMON NAMES: Skullcap, Virginian Scullcap, Helmet Flower, Mad-dog Scullcap, Hoodwort, Quaker Bonnet, Madweed.
HABITAT: USA.
PARTS USED: Aerial parts.
COLLECTION: The whole herb is collected in mid to late summer. It can be made into a brew or dried and fed as a cut or powdered herb.
ACTIONS: Sedative, nervine, anti-convulsant, tonic, antispasmodic, slightly astringent.
USE: Ideal for nervous stress or tension, and hysteria. In the past it has been found to be effective for epilepsy in both humans and dogs, and was regarded as a cure for rabies. The herb combines well with valerian.
DOSE: For a brew, add one good handful of the fresh herb or 20 grams of dried herb to ½ litre/1 pint of boiling water, and leave to infuse for 15 minutes. Give 3 cups of the brew daily. If using dried or powdered herb, mix 15-20 grams daily into feed.

> **NOTE**
>
> **Excessive doses of scullcap tincture have been found to cause giddiness and stupor in humans.** The fresh or dried herb has no such effect, therefore it is suggested that the tincture should not be used unless under veterinary supervision.

SLIPPERY ELM *Ulmus rubra, U. fulva*

COMMON NAMES: Red Elm, Moose Elm, Indian Elm, Sweet Elm.
HABITAT: Central and eastern parts of North America.
PARTS USED: Inner bark, powdered.
COLLECTION: This is usually done commercially from trees of approximately ten years of age. Unfortunately this often leads to the death of the tree, so the soft inner bark is now in short supply, and costly.
ACTIONS: Demulcent, nutritive, emollient, astringent.
USE: Rich in mucilage, slippery elm can be used as a food and a medicine. It will soothe inflammation, both internally and externally. Externally it is excellent used as a poultice (can be combined in equal proportions with powdered marshmallow) to draw poison, ulcers and boils and to encourage wounds to heal. When moistened, because of its mucilagenous content slippery elm will take on a soft and slimy consistency; this is particularly helpful when poulticing horses because the dressing will tend to stay in place whilst you are securing the bandage. Can be used internally to soothe the digestive tract, for ulceration and colitis. It is an easily assimilated food, and a specific for scouring, particularly in young foals and elderly horses. The action of slippery elm is gentle and will not harm the most sensitive of horses. When used for scouring it will reduce the inflammation present with its soothing and

mildly astringent action, rather than 'bunging' up the system, which is how the chalky conventional treatments tend to work.

DOSE: For digestive and scouring problems, mix 2 tablespoons of slippery elm with 'live' yoghurt and/or honey. For foals, slippery elm can be mixed with yoghurt and given via an oral syringe. For use as a poultice, mix the required amount with boiling water to form a soft paste and apply to the affected area (taking care to ensure that it is not too hot), cover and bandage, change the poultice regularly.

TEA TREE OIL *Melaleuca alternifolia*

HABITAT: Australia.
PARTS USED: Leaves – from which the oil is distilled.
ACTIONS: Antiseptic, mild anaesthetic, antimicrobial, antifungal.
USE: Tea tree oil products are now widely available from health food stores around the world, and the creams and ointments available are safe and effective for dealing with minor cuts, abrasions and a whole host of other conditions. More recently a number of fly repellents containing tea tree oil have come onto the market, and I have found them to be very effective. As with any essential oil, neat (undiluted) tea tree oil must never be used directly on the horse: it is very strong and if used on sensitive skin can cause blistering and flaking. Correctly diluted, however, (see details on dilutions in the Aromatherapy section) it can be used externally for any sort of fungal infection, sores, infected wounds, skin abrasions and in the treatment of ringworm.

More recently I have found the oil very useful in dealing with ticks on horses and dogs. Apply the oil directly to the body of the tick alone. (This is the **only** time I would use the oil neat.) Do this several times and once you are sure the tick is dead, remove it carefully. Tea tree oil can be used as a massage and with its mildly anaesthetic and stimulatory action can be helpful for massaging strained, bruised, damaged or rheumatic muscles. Tea tree oil can be added to water and used as a compress, or as an inhalant for horses if excess or infected mucus is present.

See section on Aromatherapy and Essential Oils, page 144.

CAUTION

Essential oils should not be given orally without professional advice. If in any doubt whatsoever, consult a qualified aromatherapist and/or vet before embarking on treatments.

THYME *Thymus vulgaris*

COMMON NAMES: Common Thyme, Garden Thyme.
HABITAT: Indigenous in Britain, Europe, northern and central Asia. Likes dry, sunny banks.
PARTS USED: Leaves and flowering tops.
COLLECTION: Collect the flowering branches when they are dry and warm. Dry the branches and then strip off the leaves.
ACTIONS: Anthelmintic, antispasmodic, expectorant, antimicrobial, astringent, carminative.
USE: The expectorant action, along with the antiseptic properties, makes thyme ideal for coughs, catarrh and chest infection. It will act as a urinary antiseptic, and is ideal for any digestive complaints, including colic. Can be used for an infected or inflamed uterus and to help expel retained afterbirth. Use externally as a brew for skin infections and irritations.
DOSE: One handful of fresh herb daily, finely chopped, preferably brewed and mixed in feed, or 15 grams of dried herb daily, brewed and mixed in feed.

> **CAUTION**
>
> **Thymol – the volatile oil in thyme – is toxic in large quantities, so use with caution. Do not use on pregnant mares. See also note on prohibited substances, page 54.**

UVA-URSI *Arctostaphylos uva-ursi*

COMMON NAMES: Bearberry, Beargrape, Hogberry, Rockberry, Mountain Cranberry, Sandberry.
HABITAT: To be found on dry rocky hills in Britain, central and northern Europe, North America.
PARTS USED: Leaves.
COLLECTION: The leaves of this evergreen plant can be collected at any time, but preferably in the spring and summer.
ACTIONS: Diuretic, astringent, urinary antiseptic.
USE: Particularly for cystitis, and infections of the urinary tract. This herb has been used with great success on mares who suffer recurrent bouts of cystitis, in combination with couch grass, marshmallow and clivers. An American herbalist told me that she uses uva-ursi to help geldings and stallions whose sheaths have become infected, or who are difficult to clean or who have an overproduction of smegma. Since then I have used this

herb successfully on my stallion and a couple of geldings. I must stress, however, that this is not a short cut to good horse management, and should only be used if the horse will not allow normal sheath cleaning or if infection is present. Uva-ursi should not be used repeatedly or on a long-term basis. It is important to get to the root of the problem and take steps to stop it recurring.

DOSE: 10 grams of the dried leaves daily.

VALERIAN *Valeriana officinalis*

HABITAT: Common in Britain, Europe, Asia, naturalised in eastern US.
PARTS USED: Roots. (There is a species of valerian native to South America where the aerial parts are used.)
COLLECTION: Collect the roots in the late autumn, clean and dry out of direct sunlight.
ACTIONS: Sedative, nervine, antispasmodic, hypotensive, carminative, laxative, febrifuge, vermifuge, warming.
USE: The name valerian is thought to derive from the Latin *valere,* meaning to be in good health or to be strong, referring either to the plant's healing properties or to its strong smell. It has had many miraculous cures attributed to it, including that of epilepsy, and was listed in both English and American pharmacopoeias up until forty years ago. It is a recognised and approved food flavouring in soft drinks and sweets in Europe and the USA. It is non-addictive.

Valerian has no connection with the drug Valium: they have no chemical similarity whatsoever; the only likeness is in the name, which one suspects may have been an adaptation of the name valerian because of Valium's sedative action.

Valerian is a common plant in Britain; it is to be found growing on walls and banks and can grow to 1.5m/5ft high.

The root has a particularly pungent aroma, which cats love (our cat rubs herself all over the sack containing the dried root). In Elizabethan times it was used by rat-catchers because of the love this animal has for the plant and its smell. Many people find the smell unpleasant, which may account for its ancient name of 'phu' or 'fu', which probably related to its less-than-pleasant aroma.

The beauty of this herb is that, in the correct dosage, it can help to relax and calm horses without them losing their faculties, becoming doped, or affecting their performance. It is ideal for relieving nervousness and restlessness, stress, anxiety, bronchial spasms, stomach cramps, colic, flatulence, constipation and nervous exhaustion. Valerian oil can be used externally as a rub for cramps and muscle tension.

Please note that the wrong diet (possibly too high a use of concentrated feeds) can have a direct effect on a horse's behaviour and if the horse is uncomfortable in its gut, it can result in restless, spooky or bad behaviour. Therefore always question the reason for the horse's behaviour and try to deal with the cause rather than 'damp down' the symptom!

DOSE: 15 grams of cut or powdered root daily. Be aware that this herb has a strong aroma and should be introduced very gradually to the horse's feed.

NB: If too much valerian is given, it can have a laxative action on horses.

VERVAIN *Verbena officinalis*

COMMON NAME: Herb of Grace.
HABITAT: By the roadside and on waste ground in southern Britain, central and southern Europe, Asia.
PARTS USED: Aerial parts.
COLLECTION: The herb should be collected just before the flowers open.
ACTIONS: Antispasmodic, nervine, tonic, sedative, hepatic.
USE: To help strengthen and restore the nervous system, particularly after illness. For any liver complaints. To promote milk production. A brew can be made to help with mouth ulcers and inflammation of the eyes.
DOSE: 1-2 handfuls of the fresh herb daily, or 20-30 grams of the dried herb daily. To make a brew, add 1 handful of herb to ½ litre/1 pint of boiling water.

Willow *Salix alba*

COMMON NAMES: White Willow, European Willow.
HABITAT: Indigenous to Britain, central and southern Europe, and widely cultivated elsewhere in the world.
PARTS USED: Foliage and bark.
COLLECTION: Harvested commercially.
ACTIONS: Analgesic, anti-inflammatory, tonic, astringent, antiseptic, febrifuge.
USE: Horses will eat the willow foliage readily if given access to it. It is ideal for rheumatism, inflammation, fevers, colic and cramp. Like meadowsweet it contains salicylic acid in the bark, which gives it its anti-inflammatory action. The modern aspirin is considered to have originated from willow and meadowsweet.
DOSE: 2-3 handfuls of the fresh leaves daily, or make a decoction with 30 grams of the bark in 1 litre/2 pints of boiling water, and give 140ml/5fl. ozs daily. The decoction can be used externally to massage sore or cramped muscles.

Witch Hazel *Hamamelis virginiana*

HABITAT: Found in damp wooded areas in North America and Canada.
PARTS USED: Leaves, twigs and bark.
COLLECTION: Harvested commercially.
ACTIONS: Astringent.
USE: Witch hazel is regarded as one of the main healing herbs by the native North American Indians. Make a decoction for external use on sore or inflamed eyes and as a compress to stop bleeding. Use externally for bruising – especially beneficial if added to the wash-down water on bruised legs and tired, sore muscles. For wounds, sores and ulcers.
DOSE: Externally, use either a decoction or distilled witch hazel, but note that the distilled product which is available commercially is not as astringent as other preparations because it does not contain tannins.

WORMWOOD *Artemisia absinthium*

COMMON NAMES: Absinthe, Green Ginger.

HABITAT: Indigenous to Britain, Europe, N. Africa, Asia, cultivated in the USA and naturalised in many other parts of the world.

PARTS USED: Leaves and flowering tops.

COLLECTION: Collect the leaves and flowering parts of the herb after flowering, approximately mid to late summer.

ACTIONS: Bitter tonic, carminative, anthelmintic, anti-inflammatory, stomachic.

USE: Wormwood is probably best known for being the main ingredient in the drink Absinthe, which was popular in France in the nineteenth century. The plant contains a green camphorated volatile oil which gave the drink its distinctive colour. The drink was subsequently banned in France in 1915 because of the damage it caused to the nervous system if consumed in large quantities. (Although it should be said that some unscrupulous manufacturers substituted copper for wormwood to achieve the green colour!) This should not, however, stop us from using the herb sensibly, in small quantities and over a short period of time. Interestingly enough, the herb is still used as a flavouring in liqueurs. Wormwood was first introduced to Britain by the Romans during the 500-year occupation. It was brought here and used to rid both the population and the cattle of worms.

Horses will eat this herb readily if they find it in their grazing, despite the fact that it is one of the most bitter herbs available. This is what makes it ideal as a digestive aid and to stimulate appetite.

As the name would suggest, it is a powerful worm expellant, especially against roundworm and threadworm. It can be used externally as an insectidal lotion for infestations such as lice, and in cases of mange.

DOSE: Make a brew using 15 grams of cut herb with 3/4 litre/1 1/2 pts of boiling water. Internally, give 140ml/5 fl ozs per day for a maximum of 3 days (it may be advisable to mix the brew with a dessertspoon of honey). Externally, use the brew as a lotion as required.

CAUTION

Do not use on pregnant mares and use only for short periods.

YARROW *Achillea millefolium*

COMMON NAMES: Old Man's Pepper, Soldiers' Woundwort, Nose Bleed, Staunchweed, Milfoil, Knights' Milfoil.

HABITAT: Common in pastures, hedgerows and waste areas in dry sunny positions. Found throughout the world in temperate climates, naturalised in North America.

PARTS USED: Whole herb.

COLLECTION: Collect the whole of the plant – stem, leaves and flowers – whilst it is flowering.

ACTIONS: Anti-inflammatory, astringent, antispasmodic, haemostatic, diuretic, peripheral vasodilator, digestive stimulant.

USE: It is said that the name 'Achillea' comes from the fact that Achilles used this herb to staunch the wounds of his soldiers on the battlefield. Yarrow was used extensively by the Crusaders, hence the military names such as Soldiers' Woundwort and Knights' Milfoil. This herb is widespread, and can be found in old pastures where it is grazed readily by horses and cattle alike. It is particularly beneficial for urinary infections such as cystitis and is a specific for fevers. It encourages the healing of wounds and burst blood vessels (epistaxis). It will improve and encourage appetite, especially if this is due to ill-health or fevers. It will improve the blood supply and circulation to peripheral blood vessels, making it a useful herb for conditions such as navicular syndrome. In some parts of the world it is used for rheumatic conditions.

DOSE: 25 grams per day of the dried herb mixed with feed. A compress for wounds can be made with an infusion of yarrow: pour a cup of boiling water over 3 teaspoonsful of the herb and allow to stand for 15 minutes.

NON-HERBAL MISCELLANY

There are a number of remedies that do not fall into the herbal category, and which can be extremely effective for horses. These can be given in conjunction with any other treatments, be they conventional or alternative. Also included in this section is advice on water and its purity, which of course is so fundamental to good health.

Cider Vinegar

There are many references to cider vinegar in this book, and I make no apology for the fact. It is something that I give to all my animals, both young and old, regardless of their state of health. I also have a teaspoon of cider vinegar in water three times a day and my parents swear by it for their arthritis.

Cider vinegar is regarded by most natural therapists as a bit of a 'cure-all', and certainly, if some of the old herbals I have read are anything to go by, it has been used for hundreds of years both internally and externally for virtually any problem you care to mention. It is particularly rich in potassium and carries over all the minerals contained in the original apple, such as phosphorus, chlorine, sodium, magnesium, calcium, sulphur, iron, silicon and many trace minerals. What I have found, both for myself and my animals, is that it is ideal for cleansing blood impurities, reducing calcification in joints and arteries, and improving overall health.

Cider vinegar can be used externally as an antiseptic on conditions such as ringworm (*see* ailments section, page 125), and as an astringent it is particularly helpful in reducing irritation in sweet itch cases – add 2 tablespoons of cider vinegar to 1 litre/2 pints of water and use as a mane and tail rinse.

A woman in Germany told me that she had a mare who had suffered from azoturia for many years, and would not respond to any treatment or change of management. She started to give the mare 50ml of cider vinegar daily on the advice of a friend and has not had any repetition of the condition since. That was two years ago.

It is important to try and obtain good quality cider vinegar, produced if possible from organically grown apples. Try and find a vinegar made from the whole apple rather than just the core and skins that are left over after the flesh has been pulped. (For recommended suppliers see Useful Addresses, page 164.)

Cider vinegar can be fed in one of two ways, either by adding it to the animal's drinking water or by pouring it over the feed. I have found both methods acceptable, but some horses are not keen on vinegar in their water. An average dose is 10 teaspoons or 50ml per day. I have to say that I tend to just 'pour in a glug' of cider vinegar rather than measure it, and my horses and dogs love it. You can add cider vinegar to drinking troughs where it can work in an almost homoeopathic way – as little as 2 tablespoons added to the drinking trough 3-4 times a week, depending on usage, will help gently dose the animals.

Honey

Wherever possible, try to use unpasteurised honey produced by native bees. Imported honeys might be cheaper, but they can come from infected hives. This will not harm you or your horse, but if you have to use the imported honeys make sure you wash out the jar before throwing it away, otherwise native bees might feed on the honey left in the jar, and carry an infection back to their hives.

Honey is a wonderful natural sweetener which can be used to entice horses to eat or to disguise bitter or unpleasant tastes. Many old-fashioned herbal remedies involved making a ball to slip down the horse's throat and these invariably were bound together with honey to cover the often bad taste and 'sweeten the pill'! The Arabs, who use herbal medicines extensively on their horses, goats and camels, give honey to their young animals.

Honey is ideal for horses with coughs: it will reduce irritation and soothe the throat. It is also reported to have a calming and sedative action.
DOSE: a tablespoonful twice a day.

Externally, honey has always been used for burns. It will reduce the painful smarting and prevents blisters forming, along with encouraging rapid healing of the skin. I have also used it to encourage hair growth on old wounds.

Probiotics

The following information was given to me by a well-respected probiotics distributor and manufacturer.

Probiotics are used to establish or re-establish the microflora that live within the horse's gut. Thus they work in the opposite way to antibiotics, which destroy bacteria. They are composed of dehydrated beneficial live bacteria.

The adult horse has many billions of bacteria, the majority of which live in the hind gut. The beneficial bacteria, or 'good bugs', break down the digestive fibre in the horse's diet, producing the B vitamins when they are needed, as only B12 can be stored in the horse's body. As well as producing vitamin K, the beneficial bacteria also top up the amino acid levels and an enzyme called phytase. Phytase enables the horse to utilise the phosphorus in its diet.

A small number of 'bad' bacteria also live in the gut, and if the right situation occurs, they will multiply rapidly and disrupt the bacterial balance. This may be evident in looseness of droppings, diarrhoea, nervousness or any combination of these symptoms.

Diet has a huge influence on the gut's bacterial balance, as do environmental factors, electrolytic balance, mental and physical stress factors. Cortisol, a steroid hormone which is released when adrenalin is produced, alters the acidity in the gut, resulting in the death of many 'good bugs', which then allows the 'bad bugs' to multiply quickly.

Probiotics can help counter the effects of cortisol and are normally used prior to a known adrenalin flow, e.g. travelling, competing, training, change of diet, change of home or yard, in fact any situation that alters the *normal daily routine of the horse*. Adrenalin is vital if horses are going to produce a good performance and if probiotics are used as suggested, the detrimental effects of the cortisol can be controlled.

A course of probiotics after antibiotic treatment, will help to re-establish the beneficial bacteria that have been destroyed by the antibiotics.

Daily use of probiotics, without a good reason, may make the gut dependent upon them (although this is not proven).

If you have any queries on the use of probiotics, contact a reputable manufacturer or veterinary surgeon who will be able to advise you on the correct dosage and application. 'A healthy gut, means a healthy horse.'

Propolis

Propolis, or 'bee glue' as it is also known, is made by bees collecting resin from plants and trees, in particular the poplar tree. To this resin the bees add essential oils and pollen, which they also collect from various plants. This substance is then worked on by the bee, who adds a salivary secretion to produce the propolis.

Bees use propolis to seal and reinforce their hives from the outside environment. They will coat any 'foreign bodies' that may stray into the hive with propolis to protect themselves from infection. Hives will be coated internally with propolis to protect the colony from outside contamination. This results in the creation of one of the most sterile environments known to nature. This wonderful substance is the bee's very own natural antibiotic, antiviral and antifungal treatment and it is possible to buy healing ointments, balms and gels enriched with this gift of Nature.

Water

In many parts of the world, the UK included, water, I am sorry to say, has become big business, and because of this it is driven by profit margins and the need to provide shareholders with healthy dividend cheques. Profitability looks set to take priority over quality, despite any assurances that the privatised water companies may give us. Water is a natural commodity whose purity and availability we can no longer take for granted. Recent findings of high nitrate levels in ground water and the recent case of aluminium pollution in the water supplies of Britain's West Country and the resulting health problems it caused, do nothing to allay our fears about the future.

Horses can be affected by water quality as easily as humans, and care must be taken to ensure that the supply is of the best quality. In the past, animals drank from streams fed by springs bubbling up from deep within the earth. These same springs are now under threat from the residues of fertilisers, silage and farm effluent which is allowed to escape and permeate down into ground water supplies. Not to mention the ever-increasing incidence of industrial and chemical pollutants entering our rivers and streams.

Well water is preferred by all animals, and I have seen both horses and cattle ignore a nearby trough of mains water in favour of a trough of well water on the far side of the field. However, even well water is not guaranteed safe, and if this is going to be used for drinking then it must be tested for heavy metal, nitrate and bacterial levels.

Water supply is yet another factor that must be taken into consideration, especially if a mystery illness presents itself, and no obvious cause can be found.

Yoghurt

Natural live yoghurt is a natural source of bacterial flora, and can be used to help re-colonise and re-balance the ecology of the gut after the use of antibiotics and their often dramatic action. Antibiotics, valuable and often vital as they are, make no distinction between good and bad bacteria; because of this they can cause an imbalance in the gut, resulting in scouring. I have often used natural yoghurt in conjunction with honey, both during and after a course of antibiotics, to help stabilise the gut in this way. Remember to use live yoghurt – the pasteurised variety contains no natural flora. Most horses will eat yoghurt quite happily, especially if combined with honey. If, however, you have a problem it can always be administered in an oral syringe.

DOSE: 2 tablespoons of yoghurt twice a day.

These days we have the benefit of probiotics (see page 66), which are a more sophisticated way of replenishing and rebalancing the gut. However, these are not always readily available and your local supermarket is more likely to stock yoghurt than probiotics!

ALTERNATIVE ANTIBIOTICS

Antibiotics, first used in the winter of 1940/41 (as penicillin), have been both the saviour and the scourge of the twentieth century. Whilst these drugs can save lives, they are now, through over-use, in danger of creating a generation with reduced immunity to commonplace bacteria, and subsequently reduced response to certain antibiotics.

The most recent concern is over the use of antibiotics in animal feeds, which is fostering the emergence of antibiotic-resistant organisms, which may be transmitted to humans through the food chain. Animals, if kept and fed correctly, would not require the constant 'drip feed' of these drugs.

Several years ago I was contacted by a very worried mare owner. Whilst her brood mare was being covered, she sustained internal injuries consisting of bruising and lacerations. These lacerations subsequently became infected, and despite being given heavy doses of an appropriate antibiotic, she failed to respond. All attempts to clear up the infection were unsuccessful and the mare's owner was told there was nothing else to be done. To add to the owner's plight, the mare had been scanned in foal, so two lives were at risk.

Plants have their own antibiotic systems, which have evolved to enable the plant to fight off attacks by fungi and bacteria. There was nothing to lose! With the agreement of the owner's vet, the horse was given a herbal 'antibiotic' mix, designed to cleanse the system and help the mare's own defences fight the infection. Within a few weeks the infection was clear, and the mare went full term with no other problems – and produced a fine healthy foal.

There are many herbs with antibiotic and antimicrobial properties. Those listed below are the ones used on this particular mare with such success.

There are obviously times when the use of conventional antibiotics are vital to safeguard life; however, their action will result in the removal of both destructive and beneficial bacteria. I have always made it a practice when using conventional antibiotics on my own animals, either orally or intravenously, to give live natural yoghurt and unpasteurised honey in conjunction with the course of drugs. (It is important that you use the live, unpasteurised yoghurt as this contains beneficial flora.) These foods will help replace the gut bacteria, thereby reducing the risk of scouring or fungal infection. If you prefer, you can use a good quality probiotic which will have a similar action. (See sections on probiotics, page 66, and yoghurt, page 68.)

Thyme contains a volatile oil which is strongly antiseptic, making it ideal for use as a wash for infected wounds, and internally for respiratory and digestive infections.

NB: Thyme could be harmful in large quantities, so use it in moderation.

Garlic. The antibiotic effect is attributed to the action of allicin, which has been shown to support the development of natural bacterial flora in the gut whilst killing pathogenic organisms. For microbial infections, garlic combines well with echinacea.

Echinacea can be used internally, externally, and in homoeopathic form. The herb is effective against both bacterial and viral attack; it is a specific for conditions such as septicaemia, and with other herbs can be used for infections throughout the body. One of the most important actions of echinacea is its capacity to stimulate the immune system, help localise infection and stop it spreading.

Rosehips have given us one of nature's richest sources of vitamin C (approximately 295mg per 100g). The hips are also a source of riboflavin, niacin, vitamins A and K, and thiamine. They will help the body recover from general debility, and strengthen its defences against infection.

Part Two

Common Ailments

Introduction

Listed in this section in alphabetical order are some of the more common problems and conditions the horse owner may encounter whilst caring for his or her horse, together with suggestions for the herbs/natural remedies that can be used to help prevent problems or treat them.

The list of recommended herbs is by no means exhaustive, and there may be several herbs available which have the same action. Readers who have their own favourites can obviously use these in preference. The herbs selected here are those that have been used successfully and safely on my own and other peoples' horses with veterinary supervision over the last decade.

In addition to the ailments, I have included useful information and handy tips that I feel may be helpful to horse owners.

For dosage and preparation of the recommended herbs see pages 15 and 17 respectively.

Remember: herbal treatments are not an exact science; they are gentle and take time to have an effect. Be patient. It took some time for the horse to develop the problem – it will take some time to solve it.

> **IMPORTANT**
>
> **It must be reiterated that this book is not to be used in place of veterinary care and expertise and as stated at the beginning of this book, no liability can be accepted by the author or publishers if self-treatment of an animal is undertaken. If the horse is not well, then the owner/handler must call a veterinary surgeon immediately to attend to the animal, and tell him or her of any medication, be it herbal or otherwise, that the horse has been given. All of the herbs suggested in this book are safe and non toxic when given as recommended; they are all on the UK's General Sales List or are a recognised food, and should not interfere with, nor reduce the effectiveness of, any conventional drugs/medication that the horse may subsequently be given.**

Common Ailments

Allergies

Skin: *See* Eczema, Dermatitis, Mud Fever, Sweet Itch.
Respiratory: For COPD, Asthma, Broken Wind, Heaves – *see* Hay Fever/Seasonal Allergies or Mucus.

Anaemia

Anaemia is caused by a reduction of red cells in the bloodstream. The red cells are those that absorb oxygen from the lungs, transport it around the bloodstream to the tissues in the body, collect the carbon dioxide and take it back to the lungs for expulsion. Anaemia can create a number of different symptoms, such as lethargy, respiratory problems, mental confusion and a lowered resistance to viral or bacterial infection. It is vital that you ascertain the cause of the anaemia: it is not correct to assume that it is just due to iron deficiency. Call in your veterinary surgeon to diagnose the cause; anaemia can be a serious condition.

Anaemia occurs when more red cells are destroyed in the liver and spleen, than are produced. This could be caused by production of red cells in the bone marrow being reduced by disease, by internal bleeding due to tumours or gastric ulceration, underactive thyroid, parasitic burden, lack of folic acid, or vitamin B12, and many other causes. Hence the reason why it is so important to establish the exact cause of the anaemia before attempting to treat it.

It is interesting to note that the following herbs and plants which are rich in iron are also rich in vitamin C, which is vital for the absorption of iron. This is just another example of how Nature creates a balance in food and herbs.

Suitable herbs are: **comfrey** (rich in B12), **kelp** (contains B12, folic acid and iron), **parsley, nettle** and **rosehips.** All are a rich source of iron.

APPETITE

INSATIABLE APPETITE: *See* Mineral/Vitamin Deficiency, Worms.

POOR APPETITE: Poor appetite can be indicative of various conditions, such as pain, stress, impaired digestion, organ damage, etc. as well as unsuitable or unpalatable food. It is therefore imperative that the cause is established before any long-term treatment is attempted.

Cases of short-term loss of appetite may be helped by tempting the horse to eat, using any or all of the following:
- Green vegetables such as, cabbage, cauliflower, runner beans, spring greens, celery, parsley.
- Root vegetables such as carrots, parsnips, swedes.
- Dried mint, pure apple juice, fenugreek seed.

Traditionally the old racing fraternity used to hang a bunch of gorse in the stable of horses that were off colour, to tempt them to eat.

ARTHRITIS

See DJD.

ASTHMA

See Hay Fever/Seasonal Allergies.

BOX REST/STALL REST

Box or stable rest is something we hear a lot more of these days. I sometimes feel that it is recommended too often. Obviously if the horse has suffered a fracture, grave injury or is recovering from surgery then it must be kept quiet and not allowed to move around, risking further injury.

However, there are occasions when I feel that if the horse were allowed daily access, even just for short periods, to safe paddocks, the mere act of moving around coupled with the resulting improvement in the horse's mental state, would accelerate the healing process.

Many horse owners run into problems when the time comes to turn the horse out, or when they try to walk a horse that has spent many weeks confined to its stable. At this point they find they are dealing with a demon on four legs – or two as the case may be! The horse is normally so over-excited and full of enthusiasm to be out again that he is difficult and even dangerous to lead, and tends to race around the paddock undoing all the good the stable rest has done.

Herbs can assist during these periods of stable rest in two ways. Firstly, the appropriate herbs can be given to help the healing process. And secondly, if the horse becomes agitated during confinement or on release, herbs can be given to help calm him. For example, I knew of a horse that was confined to his stable for many months with a fracture of the cannon bone. Initially the owner used anti-inflammatory and analgesic herbs along with comfrey to improve circulation and encourage the healing of the bone. After the inflammation and pain had gone, she continued using the comfrey, which, because of its ability to encourage cell growth, reduced the time spent in the stable by several weeks. The vet attending the horse could not believe how quickly the fracture had healed. Luckily enough this was a calm and easy-going horse and he tolerated his confinement very well.

Other horses are not so tolerant. In these cases herbs to speed healing can be used, and as the time for bringing the horse out again approaches, herbs such as **chamomile, vervain** and **valerian** will calm the horse in preparation for the big day!

Essential oils can play their part in this exercise: **lavender, chamomile** and **neroli** oils will help reduce nervous tension and anxiety. Use the oils, correctly diluted (see Aromatherapy and Essential Oils, page 144) on yourself, and on the horse when you are handling it in the stable during the confinement. Lavender oil was often used by stallion men to help calm agitated stallions. They would attach a piece of cotton cloth with a few drops of lavender oil on it to the stallion's headcollar. Bach Flower Remedies such as **Rescue Remedy** can be added to the horse's drinking water or given directly to the horse each day.

Bear in mind that the mental state of the horse can have a big impact on how quickly it heals physically. If the horse is agitated or depressed this can slow down the healing process, and in extreme cases can make the difference between whether the horse has the will to fight on against the odds. Bach Flower Remedies, in particular gentian and oak, can help in this instance.

A fascinating case history which particularly highlights the great contribution which essential oils have to offer in alternative medicine, is discussed in the Aromatherapy and Essential Oils section.

BRUISED SOLES

As with bruising (q.v.) the best remedy is **arnica**, given homoeopathically (a homoeopathic vet will advise on the dosage) as soon as you are aware of the bruised soles. If the horse will stand quietly then you can make up a bucket containing 1 litre or 2 pints of cold water, and add 10ml or 2 teaspoons of **arnica** tincture and a cupful of **witch hazel**; allow the horse to stand in the bucket or tub for 5-10 minutes two or three times a day. (Note: this treatment may soften the soles slightly, therefore care should be taken to ensure that the horse is not ridden or turned out onto stony or very rough ground.) Feed **comfrey** internally.

BRUISING

Horses, especially young unbalanced ones, are forever knocking themselves and the resulting bruises respond well to herbal treatment. The supreme herb for bruises is **arnica.** This can be used in two ways, either by giving the homoeopathic remedy internally (a homoeopathic vet will tell you the dosage), or by using an **arnica tincture** compress. Do not use arnica externally on open wounds as it may cause irritation. Use 1 teaspoonful to $1/2$ litre/1 pint of water. Soak the compress and apply to the bruised area, repeating every 4 hours. Other compresses made from **calendula, comfrey, witch hazel,** or **yarrow** are also very effective. If you have nothing else, then an infusion made with the flowers of the **common garden daisy** *Bellis perennis* (traditional name bruisewort), which is the same family as arnica, can be used as a compress. **Cabbage** or **comfrey** leaves can be bound onto a bruise. Give the herb **comfrey** internally for a few days. If the horse appears to bruise over-easily, it may be an indication of some other problem. Increasing the intake of vitamin C by using herbs such as **rosehip** or **nettle** can help, or use herbs such as **buckwheat** which contain bioflavonoids to strengthen the blood vessels.

COAT AND SKIN CONDITION

In general the internal condition of a horse has a direct bearing on the external appearance of its coat and skin. Poor function of the liver, kidneys, digestive or nervous system, along with deficiencies in diet, and even the horse's mental state, can manifest themselves in a dull, dry, scurfy, staring coat and a skin that is slow to heal. For the horse's sake you need to discover the cause of the problem and treat it accordingly.

The horse was not designed to fast, which is often recommended to help cleanse the system and improve skin condition in humans. The horse needs a good healthy diet, free from highly processed foods which may contain sugars and artificial additives. Include plenty of fresh vegetables and fruit in the diet along with a good quality hay and make sure the horse has access to plenty of good clean water (see section on water, page 67). Daily grooming is extremely important, unless the horse has been 'roughed off'. The action of the brush will not only cleanse the coat of surface dirt but will also act as a massage to the sebaceous glands under the skin's surface, stimulating and toning the skin condition. 'Finger grooming' can be done to stimulate the lymphatic vessels which lie just under the skin. In addition to this, horses generally enjoy being groomed in much the same way as we enjoy a massage. This daily routine, if carried out with care and gentleness, will not only massage and tone muscles, but also result in bringing pleasure to the horse and improving its general constitution.

Using external applications will only result in a temporary improvement and will go no way to solving the origins of the condition. The improvement must come from within. Once this has been achieved, external products can then be used to build on and enhance the results. In the same way as correct feeding can improve hoof quality, so it can also improve coat condition.

Herbs such as **burdock, nettle, clivers,** and **red clover** have an alterative or blood-purifying effect, which will support the tissues in their action of absorbing nutrients and eliminating waste products. These herbs are particularly good for ultimately improving the coat, skin, and of course the mane and tail. Herbs such as **calendula, garlic** and **echinacea** have an antimicrobial action which will help rid the horse of bacterial microorganisms that may be present on the skin. **Clivers,** being rich in silica, will encourage strong hair growth and are helpful if the horse suffers from a weak mane and tail or brittle hair. Hepatic herbs such as **burdock** and **dandelion** will support the liver and kidneys in their task of filtering waste products. As an all-round aid to skin and coat condition **kelp** is the very best thing to feed. There are now seaweed fertilisers that can be put onto the land and horses seem to bloom on them. **Cider vinegar** can be used for its alterative action internally, and for its astringent and antimicrobial action externally (*see* Sweet Itch). Rinses for horse's manes and tails can be made by infusing either **nettles, calendula, rosemary** or **chamomile**.

Castor oil is ideal for improving the condition of a dry and brittle mane and tail. Gently massage in a little of the oil to encourage hair growth and improve condition.

For external skin conditions such as ringworm, see appropriate sections.

COLIC

See also Worms.

> **IMPORTANT**
>
> **In all cases where colic is suspected, whatever form it takes, do not waste time – call a vet as soon as the symptoms appear.** The vet will be able to administer a muscle relaxant and/or painkillers. Colics can be fatal and herbs may not be fast-acting enough to save the horse if the case is severe. There is nothing to stop you giving a herbal or homoeopathic remedy whilst you are waiting for the vet to arrive, as these will not interfere with any treatment the horse may subsequently receive. In fact I have known horse owners who have taken just this action, and as a result have had the horse fit and well before the arrival of the vet. However, that is an ideal situation to be in and I reiterate: **DO NOT WAIT – CALL YOUR VET.**

The herbs recommended here are intended for prophylactic use on horses that are known to have a predisposition to colic. **Colocynthis** is a very effective homoeopathic remedy which should always be kept in the first-aid kit. The horse can then be given the remedy at the first signs of colic, and while you are waiting for the vet to arrive. It is a reliever of intestinal spasm, relaxes the gut and calms acute pain. Contact your homoeopathic vet for further details.

SPASMODIC COLIC: Spasmodic colic is the result of a spasm of the muscular wall of the intestine. The condition can be caused by a number of things including: migrating parasites, stress, nervous tension, too much rich food – notably spring grass, eating too soon after work. Some horses can colic after the stress of competition or a hard day's hunting. I know of one particularly sensitive Thoroughbred mare who is affected by atmospheric pressure. She always colics when the barometer drops dramatically just prior to a big storm.

If the problem is parasitic, then steps must be taken to deal with the worm burden. If the problem is severe then consult your vet, who may advise an intravenous wormer. Otherwise use herbs with an anthelmintic action, along with herbs that can heal and repair the walls of the gut. (See section on Worms.)

If the cause is stress-related then wherever possible try and identify and remove the horse from the cause of the stress. This, however, may be easier said than done and if this is not possible, or the cause cannot be identified, then herbs that will relax and soothe the horse and subsequently its digestive system are the most helpful. Nervine relaxants such as **chamomile** and **valerian,** both of which are antispasmodics, can reduce the tension in the gut caused by stress and anxiety.

If the problem is inflammation or irritation of the intestine, then use herbs such as **comfrey, marshmallow, slippery elm** or **liquorice.** These herbs have a soothing, demulcent action which will reduce irritation and encourage the healing of the gut wall. A combination of **marshmallow, comfrey, slippery elm, liquorice** and **valerian** has been used with great success on a number of horses who have had regular and repeated attacks of spasmodic colic over a period of years. No reason can be found for the cause of these attacks and a variety of treatments and management have proved unsuccessful. Since using the herbs none of the horses has had any further attacks, and this is over several years when further colics would have been expected based on past experience.

IMPACTIVE COLIC: Impactive colic is caused by impaction of food in the large intestine. The usual treatment is to administer painkillers to free the horse from pain whilst the impaction is cleared, and then drench the horse with liquid paraffin, linseed or salt water to stimulate gut movement. With horses who tend to suffer from this form of colic it is important that the digestive system is kept well lubricated. Use feeds with plenty of long fibre, and herbs such as **comfrey** and **marshmallow,** whose mucilagenous action will smooth the passage of the food with their slippery consistency.

COUGHS

See Hay Fever/Seasonal Allergies, Mucus.

CRACKED HEELS

See Mud Fever.

CUSHING'S DISEASE

This is a complaint that seems to have hit the headlines in the last few years. I don't know if this is because there are more cases, or whether people are now just more aware of the condition.

Cushing's Disease usually occurs as the result of a benign tumour of the

Common Ailments · Cushing's Disease

pituitary gland (this gland is the 'conductor' of the endocrine system and is responsible for producing the hormones which control the other endocrine glands in the body). The growth of this tumour causes the over-production of huge amounts of corticosteroid hormones which are released into the body. This results in symptoms such as the horse producing a long curly winter-type coat (it is often the fact that this coat is not shed in the spring and summer that alerts the owner to the problem). This is coupled with profuse sweating, progressive weight loss despite an increase in appetite for food and water, anaemia, lethargy and ultimately, in the latter stages, laminitis. As far as I am aware there is no cure. Horses can be given medication which will reduce the effects and give the horse a better quality of life for a little longer. Your vet can confirm the incidence of Cushing's Disease by testing for glucose in the urine and by carrying out certain blood tests.

In the past some success has been achieved in being able to bring some slight relief (not cure) of the symptoms to several horses with Cushing's Disease by using herbs that have an 'anti-cancer' action, along with herbs that support the liver, kidneys and glandular system:

Use **kelp** for the glandular system and as a rich source of mineral and vitamins, in which these horses can often be deficient.

Nettle can be used as a source of iron and vitamin C if anaemia is one of the presenting symptoms; it will also help with the circulation and blood cleansing, which is vital if the horse has laminitis.

Rosehips, rich in vitamin C, will help repair tissue and strengthen the horse's ability to fight infection.

Wormwood is reputed to have anti-cancerous actions. It is useful for its appetite-stimulating and tonic properties.

Burdock, excellent for liver function, has been shown to inhibit tumour growth.

Milk thistle helps to protect the liver from toxins and support liver function, especially when it is under such pressure from the imbalance of hormones in the blood.

Garlic is beneficial for its anti-infective qualities.

Clivers is useful for the lymphatic and glandular system.

Clover is traditionally used for cancerous growths, and as a general tonic where there is debilitating disease.

I do not guarantee success and certainly do not promise a cure, but the above herbs have helped several horses to maintain quality of life before the final decision needed to be made. However, I have recently heard that Cushing's Disease has been treated successfully with other forms of alternative therapy.

CYSTITIS

This is a painful and often persistent illness, though not common in horses. It occurs more frequently in mares than in stallions or geldings. In most cases it is the result of bacterial infection entering the urethra and then infecting the bladder. As in human cases, mares that have once had the infection can often show a predisposition to the condition in the future. The conventional treatment for cystitis is an extensive course of antibiotics. If this is the case then it is helpful to feed **live yoghurt** and **honey** whilst the antibiotics are being given, or to give **probiotics** to help replace the 'good' bacteria killed by the medication (see pages 68 and 66 respectively).

It is important that the horse should be kept warm and have a plentiful supply of fresh water to help flush the system. Inclusion of boiled barley water in the horse's feed will help soothe the system and this should be coupled with a cleansing diet as outlined in the section covering DJD. An excellent success rate has been achieved using this regime coupled with the use of herbal diuretics such as **dandelion,** urinary antiseptics and demulcent herbs.

Uva-ursi or **buchu,** both of which are effective urinary antiseptics, can be used in conjunction with **couch grass, marshmallow root** and **yarrow**, which are demulcent and will soothe irritation and inflammation. It is probably better in this instance to make an infusion with the herbs and add it to the drinking water. If, however, this is difficult or the horse is reluctant to drink then add the liquid or the herbs directly into the horse's feed.

DIARRHOEA

See Scouring.

DJD

DEGENERATIVE JOINT DISEASE/OSTEOARTHRITIS/ARTHRITIS

This condition can be caused by any number of things, such as poor conformation, severe sprains, the result of an accident or bad fall, repetitive stress on a particular joint, working a young horse too hard or incorrectly before its skeleton is fully developed, poor nutrition, inability to excrete blood toxins which can build up in the connective tissue of the joints, and prolonged tension of the muscles, caused by stress or skeletal defects. The main joints to be affected are: hock, fetlock, knee, pastern, and shoulder.

All these conditions begin with the breakdown of joint cartilage and inflammation of the soft tissue, followed by changes in the underlying bone, which results in pain, reluctance to flex the affected joint, swelling and lameness. It is vital that a veterinary surgeon be called to ascertain the extent of the condition, the source of the pain and if necessary take X-rays to assess any changes taking place in the joint.

Conventional treatment usually consists of using anti-inflammatory and analgesic drugs, along with steady work to accelerate the degenerative changes and speed up the fusion or stabilisation of the joints. Alternatively long periods of rest can be tried to allow the soft tissue inflammation to subside, although the lameness and inflammation generally re-occur when exercise is renewed.

It should be said that there is no 'cure' for this condition, but there are many herbs that can help. The temptation is very often to rest the horse to try and reduce the inflammation of the soft tissue; however, I am a great believer in the old adage 'use it or lose it' when applied to joint mobility in arthritis. It is vital to ensure good circulation and keep the joint moving if cartilage is to receive nutrition, further degeneration of the bone is to be avoided and mobility is to be retained. Bony spurs and narrowing of the joint spaces can occur in severely affected horses. In many cases the horse has developed the condition as the result of a mis-alignment of the spine, pelvis, etc., causing the animal to put too much strain on a particular joint. These horses can often be helped by chiropractors, osteopaths or physiotherapists who, with careful manipulation and massage, can tackle the original cause of the problem. By using herbs with cleansing, stimulatory, diuretic and anti-inflammatory action, often in conjunction with these other therapies, many horses have been able to return to a full working life and in some cases a competitive career.

My husband's horse Ryan, a big Cleveland Bay cross, went badly lame and was diagnosed at the age of six as having bone spavin of the nearside hock. The vet prescribed a proprietary anti-inflammatory and pain-killing drug, in conjunction with daily exercise. I was very loathe to give a young horse powerful drugs for an unspecified length of time, which I felt may have a detrimental effect on his liver.

After discussion with my vet, I used a combination of herbs which, as well as reducing pain and inflammation, improved circulation and helped to remove toxins. This mix was given to Ryan for approximately eight months, and the horse was worked steadily on the local roads and tracks during that period. At the end of that time, when he was no longer receiving the herbs, he was again inspected by the vet. Ryan passed a full flexion test and was given a clean bill of health. Nine years on he is still fit and well and regularly competes in local hunter trials. It is very likely that given the conventional treatment, the outcome would have been the same, but how much nicer it is not to have had to use drugs.

When it comes to treatment, the horse must be viewed as a whole and not just as a set of symptoms. The correct nutrition is vital and highly processed foods containing complex or processed sugars along with high levels of gluten should be avoided. It is important to reduce the production of toxins in the system, and a good supply of fresh vegetables such as carrots, parsnips, swede, celery, and cabbage will cleanse and support the digestive system.

Encourage the horse to drink as much as possible to help flush the system, and if he will take it add 30ml of **apple cider vinegar** to each bucket of water. If the horse will not drink the water then add 30-50ml to his daily feeds. It is important to get good quality cider vinegar, if possible organic and made from the whole apple not just the core and skin; try and find one with at least a 5% acidity level.

Herbs rich in vitamin C, such as **rosehips** and **nettle,** are important, as well as herbs that will improve the circulation and support the lymphatic system, such as **comfrey, buckwheat, nettle,** and **clivers**. Herbs with a vasodilatory action (e.g. **hawthorn**) can be used to strengthen the walls of the blood vessels and improve the blood supply to the joints. Use **celery seed** with its volatile oil to improve joint suppleness; and if inflammation is present use anti-inflammatory herbs such as **willow, meadowsweet** or **devil's claw**. Externally, oils such as **hypericum, lavender** or **rosemary** can be used to rub into affected joints or where muscle pain is present. When using essential oils remember that they must never be applied directly to the skin, and must always be diluted with base oils such as olive, walnut, almond or sunflower. (*See* Aromatherapy and Essential Oils, page 144, for details.)

Before deciding which herbs to use, you must assess exactly what help the horse needs, and then pick the appropriate herbs. We are all guilty of having got used to the 'quick fix' solution, which depresses the symptoms without tackling the causes – this is not the long-term or holistic approach that needs to be taken with herbal medicines. Be patient – the healing process takes time.

Cod liver oil is very often recommended in the treatment of arthritis in humans and it has been shown to have a beneficial effect; however, I do not personally feel comfortable with the idea of giving animal derivatives to any herbivore. Oil is excellent for horses, providing a readily available source of energy without putting additional strain on the gut, and olive, sunflower or good quality vegetable oils are all perfectly suitable.

Dehydration

This is caused by a loss or deficiency of water in the body tissues. The water levels of the body cells are maintained by a complex interaction between kidneys, hormonal regulators and electrolyte levels. This condition may be caused by loss or depletion of vital body salts, such as magnesium, sodium, potassium and calcium, through excessive sweating, diarrhoea, scouring, stress, renal failure, insufficient water intake and many other conditions. It is important that the cause of the dehydration is identified so that the correct action can be taken.

If the problem is caused by loss of body salts, electrolytes can be bought commercially and can be added to the horse's water as and when necessary. However, some horses are unwilling to take them in this form.

Nature has her own electrolytes and one of the best sources of these is the humble **dandelion.** Both its leaves and root are rich in potassium and magnesium and the plant contains sufficient quantities to replace that which the body loses. If excessive water loss, high humidity, stress and sweating are anticipated then dandelion root and/or leaf can be given before, during and after the event to help prevent dehydration. Should the dehydration be due to other factors, it is vital to determine what these are and take steps to deal with them rather than just treating the resulting dehydration.

A simple test for dehydration is to take a pinch of the horse's skin, say on the neck, between thumb and forefinger, pull it away from the body and then let go. The skin should be elastic and spring back immediately, leaving no folds. If in doubt, call your vet.

Some horses are reluctant to drink during competitions, and until recently the general feeling was that horses should not be offered water when they were hot. One of the few groups of horsemen and women who have consistently offered their horses water at all stages of competition are the long-distance and endurance riders, whose horses regularly cover distances of up to 100 miles, often over gruelling terrain.

The endurance group's success in preventing dehydration and azoturia was particularly helpful to the experts involved in preparing and caring for the horses that were taking part in the Atlanta Olympics, where high temperatures and humidity could be expected to take their toll. One of the methods the endurance riders use to tempt their horses to drink is to offer them the sweet 'sugar-beet water'.

See also Scouring, Foal Scour.

DERMATITIS

See Sweet Itch, Mud Fever/Rain Scald/Greasy Heels (US Scratches/Rain Rot), Coat and Skin Condition.

This is an inflammatory condition of the skin caused by outside agents, such as contact with irritants like detergents, acids, etc. Sometimes the solution can be as simple as changing the washing powder that the numnah, saddlecloths or rugs are washed in. Horses can have particularly sensitive skins and strong shampoos, fly repellants, hoof oils, and coat sheen products can sometimes produce some alarming reactions, ranging from blistering to raised welts on the skin. Obviously the cause of the irritation must be identified and removed. The areas affected can then be bathed with a brew made from **calendula** flowers, until the inflammation has subsided.

DIGESTIVE TONICS

FOR FEEDING PROBLEMS

For specific digestive problems, see the appropriate headings: *e.g.* Colic, Nervous Gut, Worms, Ulcers, etc.

'We are what we eat' may have become a very hackneyed phrase in recent years, but it still holds true. One cannot think of the digestive system as something that operates in isolation to the rest of the body. What the horse eats affects every other system in his body, and it must be remembered that his psychological state can also affect his digestive system and subsequently his physical condition. Be aware of your horse's attitude to his work and surroundings – he may live in the most beautiful stable, with a thick bed and be fed the best food available yet he may still be miserable because he lacks company, freedom to move, roll, relax, and interact with his own kind.

Nature has provided an abundance of herbs that can be found in both pasture and hedgerow, which will support the digestive system, promote and encourage appetite, or tempt fussy feeders. Bitters such as **horehound** and **wormwood** will stimulate the sympathetic nervous system to produce bile in the gut and encourage appetite; whilst demulcent herbs such as **comfrey, marshmallow** and **slippery elm,** are ideal if there is any acidity or inflammation present. To tempt shy feeders and promote appetite use **mint, fenugreek** or apple juice. If there is a problem with flatulence then **mint, aniseed** or even a teaspoonful of **peppermint oil** twice a day can help. If the horse is tense and reluctant to eat as a result, then consider using herbs with a relaxant action, such as **valerian, comfrey** or **chamomile**. Remember that lack of appetite or reluctance to eat can indicate more serious problems, so do not allow the situation to continue without getting your vet's advice.

If you need help in determining your horse's nutritional requirements, it is best to consult an independent qualified equine nutritionist who, after assessing the horse's workload, physical build, metabolism and regime will be able to advise you on the best feeding programme to follow.

Eczema

This term covers a wide range of skin conditions. However, when using an holistic treatment identifying the exact complaint is not so important as ascertaining its cause. Once this has been done the appropriate herb can be selected and the body brought back into balance. Like dermatitis, some eczema forms can be caused by outside irritants, so once again if this is the case, identify the irritant and remove it. Then treat the skin externally with compresses of herbs like: **comfrey, calendula,** and **witch hazel**. If the eczema is particularly itchy and inflamed then use a cold compress made from **chamomile** essential oil, which is the supreme essential oil for skin conditions, especially where the skin is inflamed and raw. (See section on Aromatherapy for details.)

Allergic reactions to feed are not uncommon in horses, so if the horse develops raised lumps, irritations or dry skin, then first check his diet. Has anything new been added that may have caused this reaction? Has his field or type of hay been changed? Are his droppings normal? The digestion may be out of balance – perhaps his liver and kidney function is not 100%. Has he been under stress recently? This can easily unbalance the gut and create problems.

To help cleanse the system of toxins use herbs such as **clivers** and **calendula**. **Garlic** and **nettle** (first make sure your horse does not have an allergy to nettle) are excellent blood cleansers. If the problem appears to be connected with the liver and kidneys, then use **burdock** and **dandelion.**

Epistaxis

Nasal Haemorrhage, Bleeding, Burst Blood Vessels

This can be a common complaint in horses, especially those involved in fast exercise. Any horse which bleeds from one or both nostrils should be examined by a veterinary surgeon as soon as possible and the cause ascertained, then treatment can be given quickly if the condition is serious. Often an endoscopic examination is carried out; this allows the vet to determine if the haemorrhage has occurred in the lungs, and if the lungs are infected or have a build-up of mucus.

The difficulty tends to be that once the horse has 'bled', the likelihood is that the already weakened blood vessels will be prone to bursting in the future. If you are to prevent a repetition of the bleeding it is absolutely vital to identify the cause of the problem. Once this has been done, appropriate herbs can be chosen to help deal with the cause and strengthen the horse's system against a recurrence. If the problem is linked with a build-up of mucus in the lungs, fungal infection or nasal irritation, appropriate herbs can be given to encourage the expulsion of the mucus, fight the infection or reduce the reaction to the irritants.

Garlic is ideal for both mucus expulsion and helping to fight fungal and bacterial infection. It is expectorant, antimicrobial and antibiotic. **Echinacea** is specific for upper respiratory tract infections. **Marshmallow**, with its demulcent action, will help protect the airways from irritants that can cause inflammation. Use **yarrow** for horses that have already 'bled', and to help prevent a reccurrence. Yarrow's specific action of staunching bleeding, strengthening blood-vessel walls and improving the blood supply to peripheral blood vessels is ideal in these situations. **Buckwheat**, with its rutin content, will help to strengthen blood vessels and ensure that the capillary walls remain flexible and elastic.

See also the sections on Mucus and Hay Fever/Seasonal Allergies for further detailed information on appropriate herbs.

EXCITABILITY

See Hyperactivity, Nervousness, Nervous Gut.

This is probably one of the biggest problems experienced by horse owners. Once again it is important to try and discover the cause of the excitement. If it is just that the horse needs more exercise or is getting too much high-energy feed for the workload, then the answer is simple. Behaviour of this sort can be caused by any number of situations: lack of experience, excitement at meeting other horses, poor discipline, fear, pain, anticipation of competition or just a rush of adrenalin caused by the atmosphere at an event. Herbs cannot and should not be used as a replacement for proper exercise, schooling, discipline, experience or to mask pain. They can, however, be used to calm a horse that gets over-excited and tense because of outside influences, 'competition nerves' or insecurity.

Herbs such as **chamomile, vervain, valerian, poppy, scullcap** and **lavender** are nervine relaxants, which will relieve the tension caused by over-excitability. Some of these herbs are also antispasmodic, so will relax the gut and affect the peripheral nerves and muscles, thereby having a relaxing effect on the whole system – without affecting the horse's senses or sedating him. Conventional sedatives tend to have a deadening effect on the nervous system, which will result in a temporary loss of the horse's faculties, inducing lethargy and stupor. Find the herb that suits your horse's character best; but remember: the reason for the excitability must be identified – it is not correct to keep the horse on these herbs indefinitely.

If the problem is 'competition nerves' it may be better to administer the herbs for just a few days before the event. Be aware of any stress caused to the horse during competition and if necessary continue the herbs for a few days after. Bach Flower Remedies and essential oils can be particularly helpful in situations such as this, for both horse and rider, as can probiotics. I have had great success in calming my own competition nerves by using the homoeopathic remedy **argent nit.**, which is for conditions such as stage fright.

Exhaustion

Horses can suffer from exhaustion for a number of reasons, such as:
- after long and continuous hard work;
- poor diet – this is a separate issue and must be addressed nutritionally;
- through stress or work requiring continuous concentration, leaving the horse mentally exhausted;
- from long periods of travelling;
- when fighting illness;
- shock;
- anaemia.

Oats are the best food for exhaustion. They are a nervine tonic which will feed and strengthen the nervous system. Rich in B vitamins, they are a specific to help horses recover after hard work. Some horses may be too tired to eat and they should be tempted by giving one of the bitter herbs such as **wormwood** to help stimulate the appetite and strengthen the digestion. A sweet tea using honey to help recoup energy can be useful. **Valerian** is one of the herbs that would appear to work in two completely opposite ways: being a sedative for nervous or anxious horses, whilst stimulating and strengthening the nervous system of an exhausted horse.

For horses that have suffered shock or have become exhausted through the stress of travelling or changing home, **Bach Flower Rescue Remedy** is invaluable. Put 10 drops of the Rescue Remedy in the horse's water bucket, or give the horse 4 drops on a piece of apple or a sugar lump. Repeat several times a day or when the bucket is refilled.

Prolonged periods of illness can exhaust a horse, as well as effect its mental psyche. In these cases it is important to support the horse with herbs that will help it resist infection, strengthen its defence systems, improve its mental state and promote a feeling of well-being.

Use herbs rich in vitamin C and minerals, such as **rosehip, nettle** and **kelp**.

To strengthen the immune system use **echinacea** and **garlic,** which will help the horse fight off infection and rebuild its resistance. **Red clover** is an excellent herb tonic for horses recovering from illness.

I spoke recently to a woman whose horse had been very ill, and on stable rest for many months. He had lost interest in even wanting to go out into the yard; he was exhausted, depressed and, in her words, seemed to have given up hope of ever getting better. He was receiving the best of veterinary care, as well as herbs to help with his illness. After discussion, it was decided to call in an aromatherapist, who tested and tried several oils before finding the ones to which he best responded. He was given two of the Bach Flower Remedies and the oils were used in a twice-daily massage. His mental attitude improved immediately and within a fortnight the horse was his normal self once again, much brighter and showing interest in his surroundings. His new positive attitude helped complete the healing process and shortly after this he was able to return to work.

EYE PROBLEMS

Injuries to eyes and eyelids are quite common in horses, and can be caused by the horse scratching the cornea on a thorn or getting a hay seed stuck in the eye. If the cause cannot be immediately identified, and if necessary the offending foreign object removed, the vet must be called as it may be an indication of something more serious.

Eyebright, as its name would suggest, is specific for any sort of eye condition, including conjunctivitis, keratitis and ulceration. The plant is both nervine and astringent and should be used internally at the rate of 20-30 grams daily. Externally it makes an excellent eye bath: make an infusion using 30 grams of herb in ½ litre/1 pint of boiling water and when cool bathe the eyes at least twice a day. If ulceration of the eye is a problem then it would be a good idea to adopt the cleansing diet mentioned in the mouth ulcer section (page 134). **Buckwheat**, with its capillary strengthening action, may be helpful if the tiny blood vessels in the eyes are damaged as a result of injury or a build-up of fluid.

Fear

Horses are designed to run from situations, people and animals that frighten them. This is their main method of defence, and they will seldom resort to physical attack unless forced to. When we stable or corral them, we remove this option which can make some horses aggressive through fear, especially if a horse has been badly treated in the past. If a horse has been subjected to cruelty or ill-treatment and is fearful of people as a result, then the road back to full trust is long and hard. Doping is not the answer because if the horse feels unable to control his own actions he will become even more frightened because he feels disadvantaged.

Nervine relaxants such as **chamomile, valerian, scullcap, rosemary** and **lavender** can help in these situations, by reducing the stress and tension the horse experiences when in close contact with humans. Helping the horse to relax will enable the handler to get closer to the horse, thus an understanding or relationship can start to develop. The process is slow and patience is needed at every turn. The Bach Flower Remedies can be particularly helpful in these situations, and essential oils can reduce tension and panic.

Horses can suffer from panic attacks: traffic, pigs and donkeys, for example, can cause a horse to take fright. Horses that have been involved in traffic accidents will lose confidence on the roads and may panic. I have found that the **Bach Flower Remedy Rock Rose** can work well on these occasions. Give the horse the remedy daily and whenever he is going to come into contact with the thing he most fears. It is often a good idea if the rider takes the remedy as well – remember: you will transmit your anxiety to the horse through your body and by doing so you will reinforce his fear.

Fertility

There are many herbs that for centuries have been used to improve fertility in animals. Their use and benefits have been passed on by word of mouth from stockman to son, mother to daughter (in the past it was normally the women who were responsible for the health of the family's animals). These herbs and their uses are steeped in folklore. They were employed either to regulate the female cycles or to improve the fertility of the male animals. It is unsure how many of them work: several of them have an oestrogenic action in animals, others work by improving the general condition of the uterus, or by toning the pelvic and uterine muscles, keeping them in a healthier condition. My feeling is that in general they work by improving the overall health and muscle tone of the animal, which in turn would effect all its physiological systems, improving libido, fitness, stabilising hormone levels and regulating its cycle.

FOR MARES: Although **raspberry** leaf will not improve fertility as such, I feel it is worth including in this section because it can have a direct effect on the condition of the uterus. Raspberry leaf affects the tone of the pelvic and uterine muscles; it will help with contractions at birth, reducing the chance of haemorrhage and help the mare to cleanse after birthing.
NOTE: Once in foal, stop giving the raspberry leaves, and then reintroduce gradually over the last 4-6 weeks of pregnancy.
Clover – Oestrogenic activity.
Fenugreek – This is a great herb for improving condition of animals, being rich in oils, protein and vitamins, including the 'fertility' vitamin E.
Liquorice – Contains oestrogenic substances, and has traditionally been used for female infertility.
Mint – Said to be an aphrodisiac and tonic for stallions.
Siberian Ginseng *(Eleutherococcus senticosus)* – This herb has been used on both humans and animals for centuries to improve sexual vitality, fertility, as a general tonic and 'cure-all'. This particular species is much less well known than the more popular **Korean Ginseng**. However, over a thousand studies have been carried out on this plant in Russia, and these have been substantiated by Western scientists more recently. One particular study carried out on bulls improved both the volume and the concentration of the ejaculated spermatazoa by between 10% and 35%.[14] It is to be hoped that as a result of this research, Western veterinary research institutes will be persuaded to carry out their own investigations into this remarkable plant. For advice on dosage, consult a veterinary herbalist.

FOALING

If the mare is a maiden or likely to become stressed during foaling, **Bach Flower Rescue Remedy** can be added to the mare's drinking water as the time of foaling approaches. Continue to dose the water during and for a few days after the birth to help reduce any anxiety or stress caused by the birth.

Once again, to help with strong contractions and cleansing of the afterbirth, **raspberry** leaves given for the last 4-6 weeks of pregnancy and for one week after the birth of the foal will tone and strengthen the uterus, assist with the healthy expulsion of the afterbirth and aid the mare to return to full health after the birthing.

Garlic given in small amounts – 3-4 fresh cloves or 10 grams of powdered garlic daily – can be helpful in cleansing the blood of impurities and helping combat infection. Be aware that using too much could taint the milk, which may put off the foal.

Foal Scour

This is quite common in foals, particularly when the mare has her first season after foaling (foaling heat) – usually 7-12 days after the birth of the foal. In the past it has been assumed that the reason the foal scours is because the mare's milk changes during the foal heat. However, recent research has shown that in fact the milk doesn't change, and it is now held that the scouring occurs as a result of worms.

If the scouring persists you must call a vet immediately as foals can dehydrate very quickly. Check for dehydration in the usual way, by pinching a piece of skin on the foal's neck and holding it tight for a few seconds. When released the skin should spring back quickly – if it doesn't, then the foal is probably dehydrated. Call the vet immediately. The liquid droppings will burn the foal's delicate skin, so make sure you wash the area thoroughly with warm water and cotton wool. Add a few drops of **hypericum** or **calendula** tincture or oil to the water to help soothe the burning, and then cover the area with olive oil or baby oil to form a protective barrier and prevent hair loss.

Mix 1 tablespoon of powdered **slippery elm** bark to a liquid paste either with some of the mare's milk or with some natural live yoghurt and syringe it into the foal's mouth. Do this two or three times a day. The **slippery elm,** which is highly nutritious, will act as an internal poultice, soothing, healing and reducing the inflammation present in the bowels. The use of slippery elm is preferable to using the conventional chalky chemical products which tend to block and 'bung up' the bowels rather than encourage them to heal.

It can be helpful to feed the mare on a little **garlic** (3 or 4 fresh cloves or 2 teaspoons of powdered garlic daily) at this time as its anthelmintic and antimicrobial properties will be passed through the milk to the foal.

Glandular Problems

The glandular or endocrine system is far too complex a subject to be covered in this book. However, like any system, it reflects the overall health and status of the body. The blood is the medium responsible for carrying the secretions of the glands to the target organ, so blood tests are needed to access hormonal levels. This, however, is not a fail proof test, as it is the condition and receptiveness of the target organ that dictates the correct and balanced response.

Underactive or overactive glands can create a confusing array of symptoms, and it is important that if the horse owner feels there is a problem, a professional diagnosis must be obtained. Once the problem has been identified herbs can be used to treat the body as a whole by strengthening it and restoring a harmonious balance.

Herbs such as the bitters like **wormwood,** and **yarrow** are best for their ability to cleanse and improve blood functions in general, treating the overall picture.

The thyroid gland which controls the body's metabolic rate can be affected by stress or mineral deficiency. Certain geographical areas may have a higher incidence of hypothyroidism (underactive) problems due to lack of iodine. Goitre, which is a human condition caused by either an under- or overactive thyroid, was also known as 'Derbyshire Neck', because of the high incidence of the condition in Derbyshire where there was a deficiency of iodine in the diet. Obviously once this connection was made, iodine could be added to the diet (often in table salt) to correct the deficiency. A veterinary surgeon based in the Channel Islands told me that she believed there was a higher incidence of horses suffering from thyroid problems in the Channel Islands because of an iodine deficiency.

The richest natural source of iodine is the seaweeds, particularly **kelp. Clivers,** which is also rich in iodine, is excellent for all glandular problems.

An overactive thyroid, which produces hyperactivity, stress and nervous anxiety, can be helped with **nettles, valerian,** and **yarrow,** which will act as hormonal tonics. **Bugleweed** *(Lycopus virginicus)* has been shown to reduce the production of the hormone TSH (thyroid stimulating hormone), which is responsible for thyroxine secretion, and is therefore a specific for hyperthyroidism. However, I have no personal experience of its use on horses. NOTE: Kelp should not be used on horses with this condition as it can be thyroid stimulating.

The parotid gland can be troublesome to horse owners. This is the gland that is situated at the base of the ear, and will often become swollen making the horse look like a hamster! It is generally believed that these glands enlarge as an allergic reaction to something the horse has eaten. I looked after a chestnut Arab gelding for many years, who if turned out to graze in one particular field would always come in looking as if he had mumps! The condition did not distress him, and after a few hours off the grass the swellings went away. I never discovered what it was that caused the allergic reaction, but I have heard from many horse owners who have experienced this problem, and it is more than a coincidence that the majority of them seem to own chestnut-coloured Arabs. I found the best way to reduce this problem was to give the horse a mixture of **clivers** and **calendula** flowers. These two herbs are the best thing for the lymphatic and glandular system. Use a 50/50 mix of the two herbs, giving 30 grams per day.

See also Hormonal Imbalances.

GREASY HEELS

DERMATOPHILOSIS

See Mud Fever (US Scratches).

HAY FEVER/SEASONAL ALLERGIES

See Sweet Itch, Head-shaking.

With the reduction in air quality and the higher incidence of chemical pesticides and crop sprays, the horse's respiratory system is under more pressure than ever before. In Britain the M25 motorway is not only notorious for its traffic jams, but has also been blamed for creating odd behavioural problems in the animals that graze alongside it. The effect that lead can have on children has already been proved, so there is little doubt that some of the horses in fields alongside motorways are being similarly affected by the pollutants that are poured into the atmosphere 24 hours a day.

Certain crops can play a key role in adding to the reduction in air quality. In Europe the recent high subsidies paid to EEC farmers to encourage them to grow oil seed crops such as rape and linseed have resulted in massive fields of blue and yellow appearing all over the countryside. These pollens are especially irritating because, although still microscopic, they are large and 'hairy' as well as having an extremely pungent smell. It is now believed that the cocktail of chemicals present in these plants may be responsible for the irritations and allergies being experienced by both humans and animals alike.

Humans have reported reactions such as sore, weeping and irritated eyes, difficulty in breathing and an increase in asthma attacks. Horse owners are finding that their horses are now suffering from breathing difficulties, spots on the muzzle, head-shaking, lethargy and severe physical distress when they are ridden or grazed near these crops. One rider told me that to reach her nearest bridlepath she had to ride her horse through a field sown with oil seed rape, and by the time the horse had reached the end of the field he was wheezing so badly he could hardly walk.

Some tree pollens such as may, oak, and beech can also create allergic reactions, preventing some horses from competing during the summer months. One horse owner in Scotland told me that whilst the may tree is in bloom she is unable to ride her horse out. If he comes into close contact with the may-blossom he tosses his head around so violently that in the past he has actually thrown himself to the ground.

All these irritants can create hyper-sensitivity in horses, and many people will find that once an allergy has been created the horse becomes more susceptible to others. Horses have been seen to stand with their muzzles partially submerged in drinking troughs to try and relieve the condition. I have had personal experience with my own horse of extreme lethargy being created by pollen allergies.

Unfortunately we have little or no control over the quality of the air we breathe. Recently 'nose veils' have become available, and these can help by filtering out some of the larger pollen particles present in the air. They are available commercially, but a piece of butter muslin or an old pair of tights/stockings can be used with equal success. Attach the muslin etc. to the noseband of the horse's bridle or headcollar and allow it either to hang down over the nostrils like a curtain if the horse is grazing or bring it round and attach it to the back of the noseband if the horse is being ridden – and, of course, if he will allow it! Make sure it is loose enough to allow the horse to open his mouth freely. Unfortunately these measures cannot be used in competition.

Homoeopathy and acupuncture can be very effective against hay fever, dust and pollen allergies, and head-shaking. Contact a homoeopathic vet or acupuncturist for advice, but if the horse is an annual sufferer remember to contact them in plenty of time. Do not wait until you already have the problem.

Humans are normally prescribed antihistamines to help them deal with hay fever and allergies, but there is a general feeling that these have a limited effect on horses. However, some herbs do have an antihistamine, anti-asthmatic and anti-allergic action, which can help reduce the reaction to pollens, dust, etc. Other herbs with their demulcent, mucilagenous and expectorant action can be extremely effective in soothing and controlling the resultant irritation, inflammation and respiratory problems.

As with any herbal treatment, if you know the horse is prone to hay fever etc., give the herbs at least 3-4 weeks before the 'danger period'.

Buckwheat has an antihistamine content, as well as containing rutin which will strengthen weakened capillaries.

Marshmallow is both demulcent, emollient and expectorant; it is excellent for any bronchial complaints and will help soothe and protect the airways from allergens.

Garlic – Being an expectorant, it will help encourage the expulsion of irritants such as dust, spores and pollen from the lungs.

Boneset – Another expectorant, and a specific for catarrh; also immuno-stimulatory.

Comfrey – Demulcent and mucilagenous comfrey will soothe and ease the airways and improve the circulation.

Aniseed – An expectorant and a specific for catarrh.

Eyebright – This is for the eyes. Make a tea with the herb and bathe the eyes twice daily if they are irritated or weeping.

HEAD-SHAKING

See Hay Fever/Seasonal Allergies.

Firstly it is important to ascertain the cause of the problem. Horses will head-shake for a number of reasons:
- **Irritation** in the nasal tracts, caused through allergic reaction to dust, pollen spores, etc. *See* Hay Fever/Seasonal Allergies.
- **Discomfort** in the mouth due to tooth problems, badly fitting or sharp bits and tack. Have the horse's teeth examined and check the fitting of the bridle and saddle. The presence of wolf teeth can create discomfort and result in head-shaking, rearing or running backwards to get away from the pain.
- **Spinal impingement or pain** – if the horse's spine is out of alignment or the nerves are pinched, this can cause sudden head-shaking or twitching. Have the horse examined by a chiropractor, osteopath or physiotherapist.
- **Stress or anger** – some riders complain that their horses head-shake in the manege but not whilst out hacking. This could be for several reasons, but it is possible that the school surface, if dry and dusty, could be irritating the horse, especially if the horse is working hard and respiration increases as a result of the extra effort. *See* Hay Fever/Seasonal Allergies for suggested herbs.

Another cause could be because the horse is put under additional mental or physical stress whilst being schooled and develops the head-shaking as an evasion or in response to muscular or nervous tension or pain. Try and evaluate the reason for the stress. Is the horse very young and being asked to perform movements it finds difficult? If this is the case it may be necessary to take a step back in its schooling. Often the use of one of the **Bach Flower Remedies** can be helpful if it is a mental rather than a physical problem. The horse may be bored and sickened by continual schooling and need a change of scenery and activity. If the problem is one of stress, then some of the herbs that have a relaxing and calming action such as **chamomile, vervain** and **valerian** could be helpful, but do not use these herbs to cover up a deeper problem.

Mineral deficiency – It has been suggested that some horses may develop head-shaking if their diet is deficient in magnesium. In very wet areas where minerals are leached from the soil, magnesium deficiency can occur in pastures. Some vets have linked this deficiency with head-shaking. This would make sense as magnesium is vital for proper muscle and nerve function. It is therefore not unreasonable to suggest that a deficiency could well result in involuntary muscle and nervous spasm.

Magnesium is available in green leafy vegetables and in such plants as carrot leaves, dandelion, marshmallow, meadowsweet, and oak. It must be remembered, however, that for the body to absorb magnesium it must have sufficient calcium supplies. Calcium is present in dandelion, meadowsweet, clivers and calcified seaweed. It is interesting to note that some of these plants contain both minerals, showing how once again Nature is in balance, and making them the ideal choice.

Hoof problems

Dry, Brittle, Slow Growth, Cracked, Poor Quality

Some horses just seem to be born with poor hooves. The theory that black horn is stronger than white is not correct; research has proved that the only difference is in the pigment granules, which do not affect the hooves' strength. A horse's hooves are far more likely to be affected by the conditions the horse is kept in, the food it eats, and the farrier who shoes and trims its feet.

The condition of the hooves can be helped by external applications: herb extracts such as **comfrey**, or essential oils (suitably diluted) such as **lavender** or **rosemary**, when applied externally to the coronary band, can help to stimulate the circulation, thereby improving horn growth. Moisture is vital; the benefit of a horse walking through dew-drenched grass cannot be underestimated. The hooves are porous and moisture is easily drawn out through the walls. In the same way as our nails need moisture to prevent them from becoming dry and brittle so too do horses' hooves.

Any herb that stimulates and improves circulation and blood supply to the foot will help improve both the quality and quantity of horn. Herbs such as **comfrey, buckwheat, hawthorn** and **nettle** will help with circulation, blood cleansing and cell growth. Herbs rich in silica, such as **clivers**, are excellent to improve horn quality – if possible, during the spring and summer months whilst the clivers is freely available, pick and give the horse several handfuls daily. **Kelp** is an excellent form of multi-vitamin and hoof supplement, containing balanced quantities of essential minerals such as iron, selenium, iodine, sulphur, copper, magnesium, zinc, as well as beta carotene, vitamins D, E, C, B1 and B2, niacin, and up to twenty different amino-acids. **Rosehip** is often recommended for hooves because it is said to be rich in biotin. Actually this is not the case: analysis of rosehips has shown them to contain only minute quantities of biotin. However, there is little doubt that long-term use of rosehips can improve hoof quality and growth. I believe the rosehip is effective for two reasons. Firstly, the massive quantities of vitamin C it contains (up to 6000mg per kg) will support and maintain healthy connective tissue and cell walls in the hoof. Secondly, that the flavonoids it contains, which include rutin, will strengthen fragile blood vessels and improve blood supply.

NOTE: Some people feed gelatine to help with hoof growth. However, since its source is the bones and hides of cattle this raises the question of whether it is right to give animal derivatives to a herbivore.

HORMONAL IMBALANCES

See also Rigs and False Rigs, and Glandular Problems.

This is rather a general term, and a clear diagnosis is required to establish exactly the nature and cause of the problem before a course of action can be taken. Ask a vet, and if necessary a chiropractor, osteopath, etc., to examine the horse to determine whether it is an hormonal or skeletal problem.

The hormonal system of the body is delicately balanced and can be affected by any number of both external and internal stimuli. Equally you must remember that injury to the pelvis, sacroiliac area or spine can often produce very similar symptoms to hormonal problems. Spinal misalignment from cranium to sacrum can be directly involved with the action of the pituitary gland. The pituitary gland is often referred to as the 'conductor' of the endocrine orchestra, because it is the gland responsible for activating hormonal production in other glands, such as the thyroid and adrenals.

Many mares that have had a bad fall in the field, in competition or damaged themselves in the stable can, years later, develop difficulties in the pelvic region, creating inflammation, stiffness, muscle tension and difficulty in performing.

In general, however, hormonal imbalances can produce any of the following conditions:
• irregular or over-lengthy seasons;
• difficulty in releasing the follicle; cystic ovaries;
• inflamed muscles in the loin region; difficulties during covering.

These problems in turn can cause personality changes; severe mood swings; cold back symptoms, dropping when mounted; inability to lengthen, extend or perform lateral movements; difficulty in rounding over jumps; nappiness; rearing; extreme sensitivity when touched; and difficulty in picking up the back feet. Some mares may behave in a very stallion-like way, offering to 'cover' other horses, rounding up the other horses in the field, and developing a very 'cresty' neck. Very often the mares could be said to be suffering from PMT or PMS. The horse can develop a Jekyll and Hyde personality, being sweet natured and easy to handle one minute, and then unpredictable and possibly dangerous the next.

In general the conventional treatment is either to suggest putting the mare in foal, to see if this will help regulate the hormones, or to administer a course of hormone treatment or implants.

To my mind the first option is only worth considering if you have already decided that you want to breed from the mare. After all, there is no guarantee at the end of it that the problem will be solved. Many mare owners have chosen this route and found that the mare settles during her pregnancy and nursing period, only for all the problems to return after the foal is weaned.

Hormonal treatment can be successful, but it can also be expensive and cannot be used indefinitely. Some of the drugs present handling difficulties, in that they must be added to the mare's food immediately prior to feeding, wasted food must be disposed of safely, and gloves must be worn when handling them. Women of child-bearing age must take particular care when handling these drugs. Implants, of course, do not have these handling difficulties.

Nature can help in these situations, not by introducing synthetically produced hormones into the system but by using herbs that have a normalising action. These herbs gently balance the endocrine glands, and in so doing revitalise and support the reproductive system.

The best herb for this, and one that has been used successfully on mares, is **agnus castus**. The fruit/seeds of this tree have the action of both stimulating and normalising the function of the pituitary gland. **It does not contain hormones** and is unlikely to result in dramatic forced changes. Clinical tests carried out in Britain would suggest that it works by interacting with the hormonal balance of the body and by gently helping the body to self-correct.

As I have said before, it is important to obtain veterinary advice as to the exact cause of the problem, at which point the treatment can be decided. If, however, there is no obvious reason for the present behaviour then the gentle action of agnus castus could be helpful. It is particularly successful with mares that appear to have PMT or PMS.

Give 15 grams between two feeds, or make an infusion with 15 grams and 1/2 litre/1 pint of boiling water, and split between two feeds/doses.

In the past mares have responded very quickly to this herb; often an improvement can be seen in a matter of days, the stress and anxiety is reduced and the mare will appear to be more at ease and 'like herself' a bit more. (*See also* case history under Agnus Castus in Materia Medica section.)

Some mares may only need the herb for a short time, after which it can be stopped and the previous behaviour will not reappear. Other mares are fine as long as they are receiving the herb, but the problems return as soon as the herb is stopped. Many mares only get these problems in the spring and early summer, in which case give the herb for this period only. I do not believe in using anything indefinitely and if the mare does not show any improvement within 6-8 weeks then she is unlikely to respond and other action may be necessary.

HYPERACTIVITY

There are many causes of hyperactive behaviour, ranging from glandular conditions to food additives and pollution. Horses, like humans, can react strongly to additives in food, and pollution or the presence of heavy metals in the atmosphere can cause hyperactive behaviour. There have been a number of incidences of hyperactive behaviour being displayed by horses who are grazed in areas known to have poor air quality. This behaviour includes irritability, aggression and an inability to rest even though physically exhausted.

As in humans, food ingredients may create these problems and horses can react strongly to particular flavourings, additives, grains, cereals, sugars and herbs! It is important if this type of behaviour occurs that a vet should be called to ascertain the cause.

If the behaviour coincides with a change of food then one should not have to look too far for the cause. If, however, this is not the case then possible irritants may need to be identified and removed by a process of elimination. Whatever the cause, once it has been identified and removed, the horse's system needs to be cleansed of the chemicals that will have accumulated as a result of the allergic reaction.

Use alterative herbs such as **red clover,** which is also a nervine relaxant. Other nervine relaxants such as **vervain,** and **oats** (rich in B vitamins), can be used along with **dandelion** and **clivers** to flush toxins from the system. A daily dose of 30ml of **cider vinegar** can be helpful in cleansing and purifying the system.

IMMUNITY PROBLEMS

The majority of the complaints I have covered so far have dealt with using herbs after the body has become weakened or affected. Herbs have a lot to offer in a preventative way, and if the body can be maintained in a fit and healthy state, it will better resist attack or injury.

Nature has a bounty of herbs that have a specific role to play in the strengthening and support of the immune system. These herbs, in conjunction with good wholesome nutritious foods, freedom to graze, the appropriate exercise and, wherever possible, a stress-free environment, can all help towards maintaining the horse in a fit and healthy condition. People who live on 'junk' food, or are under pressure at work or experiencing emotional stress in their private lives, and whose exercise is restricted, are far more likely to fall prey to any infections or 'bugs' doing the rounds (look at the so-called 'Yuppy Flu'). Why, then, should our horses be any different?

The herbs that can be used for horses in a preventative way fall into three different categories:

- **Tonics** – these are herbs which will nourish and tone the body as a whole rather than concentrate on one particular physiological system – **nettle, garlic, burdock, clivers, oats.**
- **Detoxifiers** – these are herbs that will support the horse's own elimination processes and can allow you to target areas where the horse may have a particular weakness. For example, a horse may have a propensity to lung infections or coughs because his lungs have been weakened in the past by pollution or a dusty environment. An elderly horse's liver function may need support because it has been weakened by worm or drug damage. The following herbs are suggestions depending on the area where support is needed. You may have your own favourites – remember: herbs are varied and there are always going to be several options for each problem. I have chosen these herbs because they all have a gentle action and not a strong purgative one.

  For the kidneys and urinary system – **dandelion leaf, couch grass.**
  For the liver and blood – d**andelion root, nettles, clivers**.
  For the lymphatic system – **calendula, clivers**.
  For the respiratory system – **garlic, boneset, liquorice**.
- **Immune activators** – the immune system must be looked at as a complete picture, and although there are herbs which will act as immuno-stimulants, one should use them as part of a whole package to improve the overall well-being of the horse. **Echinacea, garlic** and **calendula** are all capable of stimulating the production of white cells and strengthening the immune system. **Ginseng** is probably one of the most famous and successful immune stimulants, but its cost puts it beyond the reach of the majority of horse owners. There are veterinarians in the USA who have used products containing **Siberian ginseng** on horses with excellent results. However, I have no personal experience of this herb.

See also Post-viral Syndrome.

INFLAMMATION

In simple terms, inflammation is the body's response to injury and is either acute or chronic.

ACUTE INFLAMMATION: this is an immediate reaction by the body to any form of injury, bony changes, infection, knocks or blows. It usually produces pain, swelling, heat and a reduction of use of the injured area – for example, lameness or reduction in joint mobility. The blood vessels in the affected area dilate to enable a greater blood supply to bring white blood cells to the area. These engulf bacteria and foreign objects if infection is present, thereby allowing the healing process to commence. If this action is not successful, and healing does not occur, then **chronic inflammation** follows. Either way the same herbs or treatment are applicable.

Nature is blessed with many herbs that have both anti-inflammatory and analgesic properties. They do not act in the same way as anti-inflammatory drugs, in that instead of suppressing the inflammation (which is the body's way of creating heat and stimulating blood supply) as conventional drugs do, they aid the inflammation by strengthening and supporting those very responses. 'If you want a healthy plant then feed the soil' could be a good analogy.

Herbs: **Devil's claw** has, in research, proved to have analgesic and anti-inflammatory effects comparable with the drug phenylbutazone.[12]
NOTE: **Do not use on pregnant mares.**

Other herbs with anti-inflammatory properties are: **meadowsweet, willow, comfrey, chamomile, celery seed,** and **echinacea.**

These can be used in conjunction with herbs that will strengthen the circulation, and dilate the blood vessels: **nettle, comfrey, hawthorn,** and **buckwheat;** along with alterative herbs to cleanse the tissues and eliminate toxins: **burdock, clivers, calendula, nettle** and **echinacea.**

If inflammation is present because of muscle or bone damage, then herbs like **comfrey, clivers** and **kelp,** or fodder such as **alfalfa,** will help provide a complete range of vitamins and trace elements that the horse needs to repair muscle and bone.

Poultices can be helpful in some instances of inflammation, particularly where improved circulation to a specific area is required. **Comfrey poultices** are excellent for this; **cabbage leaves** bruised and bound onto the area are another option; and tinctures of **comfrey** and **arnica** can be diluted and used as a compress for inflammation and bruising.

Chamomile essential oil is the best for inflammatory conditions, with **lavender** as a second choice. These oils should be used in either hot or cold compresses, whichever is appropriate. (For details of essential oil compresses, see Aromatherapy section.)

Myrrh essential oil is a very valuable anti-inflammatory, and is a specific for situations where the inflammation is caused by a slow-healing or very wet wound. A good example would be an embedded blackthorn.

Homoeopathic remedies are extremely successful for inflammation and bruising – contact your homoeopathic veterinary surgeon for advice.

LAMINITIS

US – FOUNDER

This book has neither the scope, nor its author the knowledge, to advise on the treatment of laminitis; moreover, it is not unusual to find differing opinions between veterinary surgeons and farriers as to the best course of action. Therefore I will just suggest herbs which if given prior to the onset of laminitis may help to prevent it, along with herbs which can help if the horse or pony already has laminitis. The suggestions made here are to be used in conjunction with good horse management. Do not think that by giving the horse or pony a handful of herbs you can then turn it out onto lush pasture with impunity.

Laminitis, in simple terms, is the inflammation of the sensitive laminae in the foot. The blood supply to the feet is disturbed, and subsequently areas such as the laminae do not receive an adequate supply of blood. Because of the lack of blood there is a reduction in oxygen reaching the cells, which causes them to become damaged and die. Damage to the cells creates inflammation and subsequently pain and swelling which, because it is restricted within the tough hoof wall, intensifies the discomfort.

Although the most common cause of laminitis would appear to be over-consumption of rich grass, it can also be caused by shock, stress, severe concussion to the feet, a reaction to certain drugs, an hormonal imbalance (see section on Cushing's Disease) or as a result of infection, especially after foaling when retained afterbirth or haematomas and bruising can lead to a build-up of dead tissue and bacterial infection. In all cases, expert veterinary advice should be sought quickly and action taken, if necessary in conjunction with your farrier.

For horses and ponies known to be predisposed to laminitis, herbs can be given prior to and during the 'danger period'. These aim to improve the blood supply, support the lymphatic system and liver, and strengthen blood vessels.

During the spring and summer months, **nettles, clivers, comfrey** and **dandelion** can be cut and given regularly to help support the liver, kidneys and lymphatic system. Ponies and horses confined to starvation paddocks or stables will look forward to their daily ration of herbs, which also helps to break the boredom so often experienced by horses in this situation. The **clivers** and **dandelion leaves** should be eaten readily, but **nettle** and **comfrey** leaves may need to be allowed to wilt before some horses will eat them.

It is important to remember that laminitic horses and ponies on restricted diets may not be receiving an adequate mineral and vitamin supply. This can lead to mineral depletion which may result in lethargy, and the habit of eating droppings, wood or earth.

One pony I heard about had been kept on a very strict diet, which, although successful in preventing laminitis, did not include sufficient vitamins and minerals for his bodily needs. This resulted in the horse suffering from severe mineral depletion. He had drastically reduced his water intake, and in August was still sporting his winter coat; also, he was so anaemic and lethargic that he was reluctant even to walk around his paddock. His owner was only alerted to his desperate need for minerals when he managed to escape into another pony's stable, where in the space of a few minutes he completely consumed a mineral lick he found there. Once the problem had been identified a nutritionist advised on a balanced diet. In addition to this improved diet, herbs and **kelp,** rich in vitamins and minerals, were given and the horse made a full recovery over the following months.

If the horse already has laminitis then herbs such as **nettle, burdock, comfrey, clivers** and **garlic** can be given to support the circulatory and lymphatic system. **Milk thistle** will protect and support the liver; and this herb is a specific if the laminitis has been caused by drugs.

Devil's claw will help reduce inflammation and pain. Laminitic horses have been seen to seek out **hawthorn leaves** and **berries**, which will improve circulation, normalise blood pressure and dilate and strengthen the blood vessels.

Homoeopathy can be extremely successful in the treatment of laminitis, especially when a quick response is necessary. Contact your homoeopathic vet for advice.

LIGAMENT STRAIN OR DAMAGE

See under Tendon and Ligament Strain/Damage.

LIVER PROBLEMS

The liver is the largest gland in the horse's body and receives blood, rich in digested food products such as fats and vitamins. These it breaks down and stores along with carbohydrates and proteins, which it metabolises into suitable materials vital for healthy body cells. One of its other functions is to act as a filter, cleansing the blood of toxic substances, such as drugs and poisons, that may be present in the horse's body, and breaking down worn-out red blood cells. Unlike man, horses do not have a gall bladder. Because they are herbivores designed to eat almost continuously, the bile is produced in the liver and is constantly passed through the horse to aid digestion.

The liver is an incredible organ whose actions are far too diverse and complicated to explain fully here, suffice to say that it is capable of regenerating itself, taking the brunt of the infections, poisons and toxins to which the horse is subjected. If attacked by poisons, such as in the case of ragwort poisoning, the liver cells will die and be replaced by scar tissue. Fortunately the liver has the power to create new cells, which will take over the function of those that have died. However, if the damage is widespread and the new cells are not sufficient to replace the damaged ones, then liver disease can result.

LIVER FUNCTION: There are many herbs that have an hepatic action, i.e. they support and stimulate the liver, an organ which is ultimately involved directly or indirectly in all bodily functions. Because of this intimate relationship between the liver and the rest of the body, any herbal preparation that is being given as a general aid to overall health should always include one of the hepatic herbs.

Nature provides an abundance of hepatic herbs, such as **dandelion,** both leaf and root, **clivers, burdock, vervain,** and **yellow dock**, all of which strengthen, and stimulate bile production in the liver. Many of these herbs were used in the past as part of a spring tonic for animals, to help cleanse and revitalise their systems which had become sluggish after the long winter months.

LIVER DAMAGE: The liver can become damaged for a number of reasons. Liver problems in horses can be caused by any of the following: viruses, long-term drug use, infection or bacteria which cause inflammation of the liver cells, poisons or toxins which will destroy liver cells, and worm larvae which can damage the liver during migration. It is important to determine the cause of the problem and then take the appropriate action.

Any liver damage must be dealt with in the long-term, and gentle hepatic and bitter herbs such as **dandelion, burdock** and **milk thistle** are ideal. Milk thistle is excellent for any situation where the liver has suffered excessive cell damage because it speeds up cell regeneration and lowers fat deposits in the livers of animals. The active constituent, silymarin, is carried in the blood plasma, moves into the bile and then is held in the liver cells.

Remember that the liver condition has a direct effect on the horse's digestive system. Use herbs like **meadowsweet, dandelion root** and **liquorice** to support the system. Diet is very important: plenty of fresh fruit and any green vegetables the horse will eat should be given – carrots, apples, parsnips, greens, cabbage and celery are all good. If available, give freshly picked **clivers** and/or wilted **nettles.** Avoid processed cereals and fats, which will put additional strain on the liver.

Several years ago I heard from a woman in Barbados whose horse's liver function was poor, resulting in skin conditions, an inability to sweat and respiratory problems. Her first course of action was to put the horse onto a 'fruit and vegetable' diet – whatever he would eat. Several weeks later, when his condition had greatly improved, she admitted that she had been feeding him all manner of local produce, including tropical fruit!

LYMPHANGITIS

The lymphatic system is a complex one; it plays a key role in repairing damage and in removing excess fluid that may have been created by infection and subsequent inflammation. In the normal course of events this system works well and any slight swelling in the legs can usually be dispersed by normal exercise, when the action of the muscles and tendons stimulates the flow of lymph. Sometimes, however, the lymphatic vessels can become inflamed, and then the resulting swelling and pain can be difficult to deal with.

If, because of the pain and swelling, the horse is reluctant to move or put weight on the affected limb, the problem is exacerbated. Without movement the fluid and damaged tissues are not encouraged to drain, thereby causing even more pain. A vicious circle is created.

Prevention is the watch word. Herbs that will support the lymphatic, circulatory and immune systems can be used in a preventative way for horses known to have a tendency towards this problem. Herbs such as **clivers, calendula** and **kelp** will aid the lymphatic system, whilst antimicrobial herbs such as **garlic** and **echinacea** will help fight infection, and **buckwheat** or

comfrey will stimulate the circulation.

If the horse is diagnosed with lymphangitis then use herbs that will reduce the inflammation, fight infection and improve circulation.

Externally use **comfrey compresses** on the swollen limbs. These will reduce the inflammation and bruising in the area and stimulate the circulation. Internally use **clivers** and **calendula** to support the lymphatic vessels; **dandelion root, nettle** or **couch grass** to help remove excess fluid; **garlic** or **echinacea** to fight the infection; **buckwheat, hawthorn, yarrow, comfrey** or **nettle** to improve the circulation; and **devil's claw** to reduce the pain.

In conjunction with the herbs, use a cleansing diet whenever there is lymphatic trouble. Feed plenty of fresh fruit, green vegetables and carrots. Avoid diets high in proteins, processed sugars and cereals, which can aggravate the condition.

MASTITIS

This is unusual in mares and normally fairly mild. Barren mares that run milk have been known to develop mastitis. Use herbs with a cleansing and antimicrobial action such as **garlic, clivers, echinacea** and **comfrey. Apple cider vinegar** is excellent for mastitis and has been used successfully on both horses and cattle.[15] Give 25ml or 5 teaspoonful twice a day to an average 500kg/1100lb horse.

Externally you can apply a cold compress using a brew made from **dock** leaves. Use two handfuls of dock leaves with 1/2 litre/1 pint of boiling water, leave to steep until cool, then soak a piece of towelling, cotton cloth or cotton wool in the brew and bathe the inflamed teats three times a day. Alternatively, you can use **chamomile** as a fomentation.

Homoeopathic remedies are extremely successful in cases of mastitis. Contact a homoeopathic vet for advice.

MILK PRODUCTION

In the majority of mares, milk production is not a problem, and provided the mare is fed correctly during her pregnancy and whilst nursing she will have no trouble in producing enough milk for her foal.

There are many herbs that can be given to a mare that is not producing enough milk. It is important, though, that you ascertain the reason for the shortage. If, however, your vet can find no physical reason for the milk deficit then the following food and herbs may help increase milk supply.

Firstly and most importantly, a good supply of clean fresh water is absolutely vital. Succulent food is also necessary. If the foal arrives in the spring, then the fresh spring grass, rich in protein, vitamins and essential nutrients, will naturally encourage and produce a good milk supply. In addition, the exercise the mare gets moving around the pasture will help with milk production. If, however, the mare foals before the spring grass is available then succulent foods can be added to the diet to help the milk. **Lucerne (alfalfa),** with its high protein and oestrogenic action, will help, as will **oats, linseed, carrots, fresh green vegetables,** and **apples**.

Herbs such as **fenugreek** (another tri-foliate like alfalfa and clover), **nettle, aniseed, fennel, marshmallow** and **kelp** will increase the quantity of the milk, whilst herbs such as **calendula, buckwheat** and **clover** will improve the quality.

Studies on milking cows have been carried out to test the effect of cider vinegar on milk production and butterfat quantities[15]. The cows were given 1 teaspoon of cider vinegar per 50kg/100lbs bodyweight, twice a day. In all cases the quantity and the butterfat content of the milk were consistently increased. I use 5 teaspoonsful or 25ml of cider vinegar twice a day on my own broodmares during nursing and the foals have done well. It certainly is something that warrants further investigation.

NOTE

If for any reason it is necessary to stop the mare's milk production then sage or mint can be used. **Do not give sage to pregnant mares. See Materia Medica**.

MINERAL/VITAMIN DEFICIENCY

DUE TO DIET

A horse may become deficient in minerals and vitamins for a number of reasons. It is my intention in this book only to cover deficiencies caused by diet. There are many excellent books available which detail the daily requirements of vitamins and minerals for horses in various types of work. A good balanced diet, that provides all the required nutrients, can be worked out in conjunction with a nutritionist. However, herbs can be used as part of these diets and will not only provide an excellent natural and bio-available source of vital minerals and vitamins, but also have the added advantage of their 'medicinal' action as well.

Few horse owners are lucky enough to own their own grazing land, and as such are not in control of their own pasture management. Paddocks can quickly become horsesick, and if the land is either heavy clay or sandy then minerals can easily leach from the soil making the paddocks mineral deficient.

Certain regions can suffer from mineral 'lock-up'. This results in the minerals not being released into the plants but remaining locked up in the soil. For example, there are areas on the Somerset Levels in England where, due to high molybdenum content, magnesium is not released into the grazing. Cattle grazed on this land have to be supplemented with magnesium otherwise they can suffer a condition called 'staggers', which can be fatal.

Ideally a soil sample should be taken and then a suitable dressing – organic fertiliser or seaweed – applied to redress the balance. This, however, is not possible if you do not have control of your own land. In these circumstances the use of herbs rich in minerals and vitamins can be helpful.

Behaviour such as the eating of soil, wood or droppings can indicate a mineral or vitamin deficiency. In cases where there is severe mineral lock-up in grazing land, horses can develop what appears to be an insatiable appetite. (See section on laminitis.)

I was contacted by a man whose young liver-chestnut horse was losing pigmentation in his coat and flecks of white hair were appearing on the body. The skin in the area around the horse's eyes had become very black in colour, in fact the horse appeared to be wearing spectacles. The owner noticed that this condition always occurred after the grass at his livery yard had become eaten down in the summer, never in winter when the horses were stabled and fed hay bought in from another area of the country. The land the horse was kept on was sandy and in an area known to suffer mineral deficiency. The owner felt that the symptoms were very similar to those shown by cattle suffering from copper deficiency. A herb mixture rich in copper and sulphur, and which included kelp, was given to the horse. Within a few weeks the colour had returned to the horse's coat and the black be-spectacled appearance had disappeared.

I heard of another similar case from a horse owner in the north of England. His liver-chestnut colt had always had a rich deep colour whilst he was living out on the moors. On bringing the horse into his stable yard to start handling and schooling he noticed after a couple of months that his coat colour had started to fade and become wishy-washy. Having had exactly the same problem with the colt's father, he made the same connection as I had done and added a balanced mineral supplement to the horse's diet. Within a couple of weeks the colour had returned to the colt's coat.

The best all-round mineral source for horse owners is **kelp.** It can either be applied as a soil dressing (see list of suppliers, page 165) or can be given in

Common Ailments · Mineral/Vitamin Deficiency

the feed. Use a good quality **deep-sea kelp** harvested from clean waters. There are forty-six minerals present in kelp, including iodine, potassium, sodium, selenium, calcium, magnesium, manganese, iron, cobalt and copper. Kelp also contains a wide range of vitamins.

Below is a list of the minerals and vitamins most important to horses and the plants and food they can be found in. Remember that soil type, weather during the growing season, and time of harvesting will effect the mineral and vitamin content of these plants.

Calcium: clivers, dandelion, meadowsweet, chamomile, watercress and willow, high levels in calcified seaweed (feed as kelp), fenugreek, cider vinegar.
Copper: burdock, clivers, dandelion, garlic, yarrow, carrots, turnips.
Iodine: high levels in kelp, clivers, garlic, carrots, cabbage, peas.
Iron: kelp, parsley, nettle, comfrey, dandelion, hawthorn, raspberry, vervain, watercress, couch grass, cider vinegar.
Magnesium: kelp, dandelion, clover, cider vinegar.
Manganese: kelp, carrots, turnips, oats, nettle.
Phosphorus: kelp, fenugreek, golden rod, liquorice, calendula, meadowsweet, watercress, carrots, cider vinegar.
Potassium: dandelion, comfrey, couch grass, meadowsweet, nettle, kelp, cider vinegar, carrots, turnips, parsnips, apples.
Selenium: kelp.
Sulphur: garlic, calendula, meadowsweet, nettle, kelp, cider vinegar.
Zinc: kelp.
Vitamin A: comfrey, dandelion, kelp, carrots, greens, nettle, parsley, watercress, fenugreek, rosehips, couch grass.
Vitamin B: garlic, oats, fenugreek, dandelion, kelp, couch grass, rosehips.
Vitamin B12: comfrey, kelp.
Vitamin C: nettle, rosehips, kelp, fenugreek, dandelion, garlic, meadowsweet, watercress, parsley, fresh greens.
Vitamin D: kelp, dandelion.
Vitamin E: kelp, fenugreek, watercress.
Vitamin K: rosehips.
Thiamine: rosehips.
Riboflavin: rosehips.

I think it is interesting to note how both the kelp and cider vinegar crop up again and again in these listings. You would not go far wrong if both of these were included in your horse's daily ration.

Mood Swings

See Hormonal Imbalances.

Mucus

The presence of mucus is a natural thing for horses, since their airways are lined with a thin film of mucus. This mucus traps small particles of dust and debris on its surface, which is then moved up and out of the lungs into the horse's throat by the constant beating of small hairlike projections called cilia. The horse then swallows the mucus.

A slight, watery mucous discharge from both nostrils is quite normal, especially after the horse has been exercised. If the discharge becomes thick, foul-smelling or coloured, especially if coupled with coughing, then this is an indication that something is wrong. Have the horse examined by a vet to ascertain the cause of the problem; sometimes an endoscopic examination is necessary to confirm the source of the discharge.

The way a horse is kept will have a direct bearing on the condition of its respiratory system. Check that the bedding is clean and dust-free, that the hay is not dusty or mildewed, and that the horse has a good supply of fresh air. A big improvement in the horse's condition can very often be brought about just by changing the type of bedding, steaming or soaking the hay, or using one of the proprietary dust-free vacuum-packed 'haylage' type products, and allowing the horse as much access as possible to fresh air and exercise. Try to feed the horse its hay and hard feed from the floor. In this way the horse is encouraged to put its head down as much as possible, which will help drain any discharge. With regard to diet, some cereals such as barley can encourage the production of mucus so these should be avoided whilst the condition is present.

The discharge may be being produced in response to other irritants – many horses will develop mucus through an allergic reaction to pollen or dust during the spring and summer months. (See section on Hay Fever/Seasonal Allergies.)

Nature has provided us with a generous supply of herbs that have expectorant, antimicrobial and demulcent actions. These will help to strengthen the expulsion of the debris-laden mucus, attack any infection present, and soothe any irritation in the airways.

Garlic is probably the best-known herb for the respiratory system. It is a recognised remedy for coughs and catarrh, and with its expectorant action will help rid the airways of excessive and infected mucus. It has been used for this purpose on both man and animals for the last 5000 years. The antibacterial compounds in **garlic** are excreted by the mucous membranes in the lungs, which accounts for its use in all respiratory conditions. **Echinacea**, with its immuno-stimulatory, antiviral and antibacterial actions, can also be used for any respiratory conditions, especially where viral or bacterial infection is present. It is a herb that can be used safely in conjunction with antibiotics if these have been prescribed.

Aniseed has long been used for coughs and its pleasant flavour makes it a favourite with horses. It has been found to increase the movement of the cilia, thereby accounting for its expectorant action.

It is the high mucilage content that gives the **marshmallow** leaf and root its demulcent qualities. It is ideal, especially when used in conjunction with **liquorice,** for soothing the throat and for irritated and inflamed mucous membranes.

I have found **cider vinegar,** with its antibacterial action, to be useful. Add 25ml to the feed twice a day or add to the drinking water.

Eucalyptus, tea tree or **lavender** essential oils can be used as inhalants to encourage the expulsion of mucus and to clear the airways. It is worth noting that **tea tree** oil has a stimulatory effect on the senses and is therefore better used in the earlier part of the day, rather than later when the horse needs to rest and relax. **Lavender** is best used for the latter part of the day, as it has a relaxing and soothing effect on the senses. Remember that essential oils are powerful and should never be used neat on the skin, particularly on the very delicate skin around the nostrils. Always dilute the oils with a carrier oil or in hot water. (See the section of Aromatherapy and Essential Oils, page 144.)

There are now nebulising diffusers available on the market which can be used to disperse essential oil droplets into the stable atmosphere, helping combat airborne infections and improving the general environment. Care must obviously be taken when using these machines to ensure that they are safely positioned, as high as possible and out of reach of the horse. They are quite costly so possibly they would only suit large yards where they may be used repeatedly.

Alternatively try one of these methods:

1. Make a pad of cotton or cotton wool and put a few drops of the chosen oil onto it. Hold the pad close to but not touching the horse's nose, for a couple of minutes, 2-3 times a day.

2. Make a pad as above, dot with oil and place it in the base of an old-fashioned 'nose bag' or 'bucket' muzzle, taking care to ensure there is plenty of air supply, so that the oil vapours can be inhaled but the oil does not come into contact with the nose. Leave on for 5-10 minutes, twice a day.

3. Add 10 drops of eucalyptus or tea tree oil to a cup of boiling water, pour over fresh bran and use in a nose bag. Allow the horse to inhale for 5-10 minutes, twice a day.

Mud Fever/Rain Scald/Greasy Heels

US – Scratches and Rain Rot

Mud fever, rain scald and greasy heels are all distinct conditions. They are all, however, caused by the same organism – *Dermatophilus congolensis* – which attacks the surface of the skin, making it raw and producing infected scabs. Although occurring in different areas on the horse, and sometimes at different times of the year, as far as herbal treatment is concerned *they are all dealt with in the same way*.

Any type of horse can suffer from rain scald, but mud fever and greasy heels tend to be found in heavy horses and cobs that carry a lot of feather. However, any horse, particularly those with white socks or sensitive skin, can, if exposed to prolonged periods of mild wet conditions, become infected. Good management is vital, and each day the horse should be checked for any signs of cracked skin, tufted hair or infection, wherever it appears on the body.

The first step must be to remove the horse from the wet conditions, take off any heavy feathering so that the infected area is easier to keep clean and dry, wash with a mild soap, removing any scabs as you do so, and thoroughly dry the area. Then treat both internally and externally with herbs that will fight the infection.

INTERNALLY: Prevention is better than a cure. If you know the horse is predisposed to these problems, then for at least four weeks prior to, and throughout the critical period use herbs that will stimulate the production of white corpuscles (these do most of the defensive work), help cleanse the body of toxins, strengthen the immune system, fight the infection, cleanse the blood and strengthen the coat and skin. Herbs rich in sulphur are particularly helpful with this condition. Herbs: **clivers, calendula, garlic, echinacea, seaweed, thyme, meadowsweet, nettle.**

EXTERNALLY: It is important to cleanse the whole area. You can then apply a poultice to draw any pus that is present. Use **clivers** mixed with **slippery elm** or powdered **marshmallow** root for the poultice (see section on how to make a poultice, page 20). Poultice first thing in the morning, pull an old stocking or long sock over the hoof and up to just below the knee, and then bandage over to keep the poultice in place. In the evening remove the poultice, clean and dry the area again and re-poultice if necessary. If the condition is not bad enough to warrant poulticing then use creams or ointments containing any or a mixture of the following: **garlic oil, calendula extract, comfrey, hypericum, tea tree oil,** or **propolis**. Propolis is a resin collected by bees from trees and used in the construction of their hives. It is the bees' very own and very efficient antibiotic. (*See* Non-Herbal Miscellany section).

IMPORTANT NOTE:
I was alerted to an interesting case of 'mud fever' several years ago by a well-respected equine nutritionist, Gillian McCarthy, in her book *Pasture Management for Horses and Ponies* (an excellent reference book for any horse owner – see Bibliography). In it she cites an instance of a dark chestnut mare with two white socks and a white blaze, who was suffering from extensive hair loss. She had scabs exuding pus on her face and on her white fetlocks. She had been diagnosed as having mud fever and had been treated with cortisone cream for weeks with no effect. She was being fed on alfalfa/timothy hay at the time, and had been wormed with a thiabendazole wormer. Within days of stopping the hay and replacing it with a clover-free meadow hay the condition started to clear up. Some weeks later she was inadvertently given the wrong hay and her legs immediately 'blew up', taking fourteen days to return to normal size.

Gillian states in her book that this is not a common problem. However, it is worth bearing this case in mind if your horse matches these details. In the past five years I have been told of three instances of this condition and the horses involved were either chestnuts or greys who had been grazed on clover-rich pasture or given clover/alfalfa-content fodder. All three cases had been diagnosed as having mud fever and had been conventionally treated with little or no success. One little grey pony who had been turned out onto clover-rich pasture had so many pustules erupt around the coronary band that it deformed the hoof growth in the following months. Once they had been taken off the pasture or hay all three horses responded within a week. All the horses were given **clivers** and **calendula** to help cleanse the system.

MUSCLE DAMAGE/WASTAGE/TENSION

Horses are athletes, and it is only natural that in the course of their lives there will be occasions when, through over-exertion, accidents, falls or just high spirits, they will suffer from muscle damage, strain, wastage or tension.

There are many other reasons for muscle problems. Incorrectly fitting saddles can restrict blood supply or nerve stimulation, resulting in uneven muscle development or wastage. Tension and inflammation can flare up in muscles adjacent to areas of skeletal injury. Muscles can become atrophied if there is nerve damage. Muscle cells can break down in conditions such as azoturia. Incorrect schooling can create tension.

Horses may suffer from rheumatism as a result of toxins building up in the tissues; even stress and anxiety can create muscle tensions which can lead to lameness and inflammation. We all know in our own lives how tension can create restricted movement in the neck and shoulders, leading to muscular pain, headaches and incorrect posture. Killing the pain is not the complete answer, although for the horse's sake if it can be dealt with in conjunction with addressing the origins of the problem, so much the better. Very often, by removing the pain the horse is better able to respond to other treatment.

The important thing is to identify the original cause of the problem. Once you have done this you can then take the appropriate action, be it calling in a chiropractor, a saddler, changing your way of schooling or training, or removing the cause of the stress. Whatever the reason, herbs can help by dealing with the horse's system as a whole. In all of the aforementioned conditions the body will have accumulated toxins in the affected area, so treat the body by supporting it and helping it to rid itself of them. A cleansing diet, as discussed in the section on liver damage (q.v.), should be used along with the herbs appropriate to the condition.

To cleanse the blood and help remove any toxins present at the site of injury, use the alterative herbs such as **celery seed, clivers, burdock, kelp** or **nettle.** If you feel that inflammation is present then use anti-inflammatory herbs which will support the body in its natural action to reduce inflammation, such as **meadowsweet, chamomile, devil's claw** or **calendula.** However, you must not use these in isolation: remember, inflammation is an indication of something wrong such as damage, infection or toxic build-up. It is particularly important to improve circulation in cases of muscle damage or wastage: **rosemary, buckwheat, comfrey, nettle,** or **hawthorn** will improve blood flow to the affected muscles and help cleanse blood toxins.

The liver and kidneys are under particular pressure on these occasions because it is their role to cleanse the system by filtering and eliminating waste products. Use diuretic and hepatic herbs such as **dandelion leaf, nettle, yarrow, couch grass,** or **celery seed** to help the kidneys, and **dandelion root, milk thistle** or **meadowsweet** to help the liver. For the lymphatic system, which is responsible for tissue cleansing, use **calendula, clivers** or **fenugreek.**

Externally, herbs or oils can be used on the muscles to stimulate the circulation in the area, thereby increasing blood supply, which will relieve inflammation. On areas where there is pain **hypericum** oil (St John's wort) can be massaged in gently.

Essential oils such as **lavender, chamomile** and **rosemary,** correctly diluted with a carrier oil, can be massaged directly onto the affected area. A warm compress of **rosemary** or **thyme** can be used. Make an infusion with 30 grams and 1/2 litre/1 pint of boiling water. Soak the compress in the infusion and apply as hot as possible (taking care that it is not too hot), then bandage in place and repeat every 4 hours. If the skin is not broken in the area then **arnica** tincture is excellent for sore, bruised or inflamed muscles: add 5ml/1 teaspoonful of tincture to ½ litre/1 pint of hot water, soak the compress and apply as warm as possible (taking care that it is not too hot), then bandage in place and repeat 2-3 times a day if possible. In the past I have made a lotion using 5ml arnica tincture, 5ml comfrey extract and 5ml lavender oil mixed with 500ml of aqueous cream and massaged it into affected muscles. This is a particularly good lotion to use if the area is difficult to compress and it can be used both before and after manipulation. The **arnica** and **comfrey** will help with any bruising and inflammation, whilst the **lavender** oil will stimulate the circulation and help disperse the toxins which build up in sites of injury.

The homoeopathic remedy arnica can be used for muscle injury and there are now a huge selection of treatments available, including magnets, Faradic, 'H' wave, laser and acupuncture, which are extremely successful especially if used in conjunction with an internal herbal treatment.

Navicular Syndrome

This is a condition which seems to be on the increase, or maybe we are just more aware of it these days. It can affect any age and any breed of horse, though it is not usually found in ponies. People talk about classic navicular feet as being broad, long in the toe with collapsed, dropped or contracted heels. However, horses with short, upright, narrow feet can also suffer from navicular. The important thing is to have the horse carefully examined and professionally diagnosed. At this point you can start to take steps to deal with the condition.

There are many treatments and one of the most important is the correct trimming of the foot: get yourself a really good farrier who can offer corrective shoeing along with the long, slow process of improving the foot shape and angle. The horse must be allowed to move around freely if a good blood supply to the feet is to be maintained. Try not to have the horse standing in the stable for any longer than is absolutely necessary.

Conventional veterinary treatment usually involves the use of analgesics, anti-inflammatories, vasodilatory drugs which will dilate the peripheral blood vessels providing a better blood supply to the affected area, blood-thinning drugs such as Warfarin, along with more dramatic treatment such as nerve desensitisation (this last option creates temporary relief, but the nerves do regenerate). Some of these treatments are successful, especially if the condition is diagnosed in its early stages and can produce temporary and even permanent soundness.

Once again the whole body must be viewed rather than an isolated area. It is generally accepted that navicular syndrome can be caused by an impaired blood flow in the back of the foot. This could be due to conformation, thickening of blood-vessel walls, blood-clotting, or changes in the navicular bone caused by poor or altered blood supply. There is still a great deal that even the experts do not know about this condition. Suffice to say that I have known of many horses that have been written off, coming back to a full and productive life after using herbal remedies.

The holistic approach does not differentiate between ailments such as arthritis, rheumatism, muscle damage, navicular, etc. They are all treated in the same way herbally. By viewing the body as a whole, and by supporting and strengthening all its systems, we can return it to full health. For herbs and dietary suggestions, follow the recommendations given for DJD, for although these conditions are not the same, the herbal approach is.

In addition particular success has been seen using **hawthorn** and **buckwheat.** These plants have the action of both strengthening and dilating the peripheral blood vessels, without affecting the blood pressure. **Buckwheat** is a specific if blood-clotting is suspected. **Devil's claw** can be used to reduce inflammation and pain, without creating a false impression which may encourage the horse to over-use himself. **Cider vinegar**, 25-30ml twice daily, is a must. This wonderful product has produced excellent results.

Nervous Gut

The gut is easily affected by the mental condition of the horse. Stress-inducing situations, such as a change of home, travelling, competing or difficult schooling, can all create nervous tension which can duly affect the digestive system. Problems such as colic, malabsorption of food nutrients, weight loss, scouring and the subsequent poor condition can be the end result of a nervous or tense horse. They can also be an indication of more serious problems, and it is important to identify the root cause. What we feed the horse is equally important, and sometimes an allergic reaction to certain foods can create malabsorption, irritation and inflammation.

If it is felt that the gut bacteria balance has been affected through stress, then a good quality **probiotic** can be particularly useful. (For further information on probiotics see Non-Herbal Miscellany section.) Many competitors give a probiotic both before and after travelling or competing, to help support and rebalance the gut bacteria levels. Natural yoghurt with honey can be used along with **meadowsweet**, which will settle the gut and reduce excess acidity. Nervine relaxants and antispasmodics such as **chamomile** and **valerian** are ideal if nervous tension is suspected. If scouring and malabsorption are the problem then use herbs with a demulcent, soothing and healing action, such as **marshmallow, comfrey** and **liquorice**. These will all help to reduce irritation in the gut. **Slippery elm** and **meadowsweet** are specifics for scouring. They will reduce the inflammation and irritation of the intestinal lining and bowel that causes diarrhoea. Very often gut tension can cause flatulence. If this is the case then using **mint** or **aniseed** will help.

All the above herbs are designed to help once the problem has been identified. The ultimate aim must be to remove the original cause of the tension rather than respond to the resulting condition.

NERVOUSNESS

See Excitability, Fear, Nervous Gut.

The section covering excitability (q.v.) gives full details on suggested herbs, oils and Bach Flower Remedies that can be given. The important thing to do is to try and discover the reason for the nervousness in the horse. Is it as a result of bad handling, inexperience or previous cruel treatment? With previously unhandled young horses lack of contact with humans can create great nervousness and fear. Smells and certain objects or vermin in stables can often create unreasonable reactions.

Every care must be taken to try and reassure the horse and instil confidence. There are some event riders who say their horses would jump anything for them, simply because the horse has supreme confidence in its rider.

I have used **lavender** oil on both myself and my horses to help reduce nervousness. Bach Flower Remedies such as the **Rescue Remedy, mimulus** and **rock rose** can be helpful for both horse and handler.

OLD AGE

The majority of horses, given a healthy balanced diet, regular exercise and care, can live to a ripe old age. Native ponies and less highly bred horses can easily reach the age of thirty and remain healthy, active and have a good quality of life. The mistake we so often make is to 'retire' them. Horses that have had human contact every day of their lives, are suddenly dumped in a field with nothing to occupy them. These horses have had an active life and they do not understand why they are no longer being ridden or used each day. They can become depressed and lose interest in life. I have heard of many horses who, although they could not be ridden any longer, have been broken to harness at the age of twenty and have had many more useful and enjoyable years of life. These horses can help with simple duties in grass management, such as pulling a light chain harrow to help keep the paddocks and schooling areas in good condition. They make excellent schoolmasters and are ideal to give confidence to younger inexperienced horses. I know of one nineteen-year-old Hackney pony who, since retiring, has been taught 'airs above the ground', and piaffe and passage in long reins!

Like humans, as horses get older their bodily systems may not be as efficient as they once were; organs that have worked hard may need some additional support. The gut may not be as effective at extracting vital nutrients, or be more easily affected by changes in grass and hay supply; parasitic damage may have taken its toll. Horses may become stiff and their legs swell overnight if confined to stables. Problems such as rheumatism and arthritis may appear (see relevant sections), and they may have difficulty in chewing food correctly because of tooth problems. With a little care and preventative treatment the majority of these problems can be overcome. Obviously a thorough check-up is helpful, and if necessary an equine dentist can look at the teeth, especially if the horse is wasting food or quidding. The diet may also need attention, as an older horse's daily requirements will differ from those of a horse that is working hard.

Use herbs that have an alterative action; they will gently cleanse and restore the proper function of the horse's body. Suggested herbs include **celery seed** (also very good for rheumatism), **clivers, burdock** and **kelp**. Older horses' circulation can be helped by using **nettle, hawthorn, rosemary, buckwheat** and **comfrey**. These herbs will help cleanse the toxins by increasing the blood flow.

Cider vinegar is excellent for use on animals of all ages, and if given from an early age can often help to prevent the occurrence of conditions such as rheumatism and arthritis later in a horse's life.

Some elderly horses can suffer from scouring, especially if there is a sudden change from grass to hay or vice versa. Many horse owners experience problems when bringing horses in during the winter months. The horse has shown no problems whilst out at grass, but starts to scour within a week of being brought in off the grass and fed on hay. My personal feeling on this is that the gut is not prepared for so dramatic a change in diet. I have found this situation can be avoided by introducing the hay whilst the horse is still out at grass, to help accustom the gut to the change in fibre source. **Alfalfa** can often be used as a go-between in these cases. **Probiotics** can help replace and recolonise the gut with 'good' bacteria if the balance has been upset, and in extreme cases when the scouring persists the use of demulcent, nutritive and soothing herbs, such as **slippery elm, marshmallow** and **liquorice**, can be employed to good effect.

PMT/PMS

See Hormonal Imbalances.

Pigmentation Loss

Mineral Related

See case history in Mineral/Vitamin Deficiency.

Post-viral Syndrome

This is a condition that we associate more with our top athletes, than with horses. However, horses are athletes too, and recently they have been displaying similar symptoms to their human counterparts. It would seem that, like human athletes, the horse is most at risk when fully fit, competing and under stress. The greatest incidence of this condition occurs in the racing world, where horses come into close contact with each other before, during and after the race. The horse can appear perfectly fit and healthy, and often the trainer's first indication that something is amiss is when the horse's performance deteriorates, by which time the damage is done.

Racing is a stressful environment at the best of times and if the horse is raced whilst trying to combat viral or bacterial infections the immune system may find it just too much to cope with. In addition to the drop in performance, the horse may burst a blood vessel and 'bleed', become lethargic, lose enthusiasm for its work, cough and have a mucous discharge. Blood tests taken at this time normally show a high white cell count, confirming that the horse's immune system has been activated to fight off infection. The usual course of action is to administer antibiotics, steroids and expectorant products, and take the horse out of work. Unfortunately in the commercially driven world of the Turf, the trainer is often under pressure to bring the horse back into work too quickly. On the face of it, the horse appears to have recovered; the blood profiles are back to near normal and the horse is returned to work. Unfortunately in some cases, within a short space of time the condition reappears and the whole scenario is repeated. This situation can continue, with some horses never returning to full fitness.

To a far lesser degree this condition can affect horses working in other competitive spheres, such as eventing and show jumping. Fortunately, in most cases the owners are not under the same commercial pressure as the race trainer, and the horse is more likely to be given sufficient time to recover fully.

This is the key: the horse must be let down gently and given time to rest and recuperate. At the same time a cleansing and health-giving diet, low in acids and supplemented with plenty of fresh fruit and vegetables, should be fed in conjunction with alterative and hepatic herbs. These could include **burdock, dandelion** and **clivers,** which will support the liver and kidneys in their task of cleansing the horse's system of waste products and toxins.

Herbs such as **rosehip, nettle,** and **kelp**, that are rich in vitamin C and minerals, should be fed to help strengthen the horse's natural defences. Antimicrobial herbs such as **echinacea, calendula** and **garlic** can be given to help fight infection.

A couple of years ago I was in contact with a woman whose show jumper was suffering from this condition. The horse had been taken out of work, rested for several months and then gradually re-introduced to work. But as soon as a little more was asked of the horse, all the old symptoms reappeared. This had happened on several occasions over a period of just under two years, and the owner was at her wits' end as to how to return this promising young horse to full health. As luck would have it, the owner had a friend who was a vet. This meant that she was able to get regular blood samples tested to enable her to assess the horse's improvement. The horse was given the herb echinacea, which recent research has proved is both antimicrobial and effective in stimulating and strengthening the immune system. The herb was given at the rate of 10 grams daily, and blood tests were taken every two weeks. Within six weeks a marked improvement was noticed in both the blood profiles and the horse's general demeanour. The horse continued to improve and after less than four months on the echinacea the horse was pronounced fit. The horse returned to full fitness a short time after this and has continued in good health ever since.

Rain Scald

US Rain Rot

See Mud Fever.

Rheumatism

See also Muscle Damage/Wastage/Tension.

When using herbal remedies it is important to remember to treat the whole body. If symptoms are dealt with in isolation, any improvement is likely to be only temporary. This is particularly true when dealing with rheumatism, where the resulting pain, possible lameness and or muscle wastage should not be treated as separate issues.

The aim is to restore and support the horse's body as a whole, allowing it to come back to full health. Horses, like any living creature, will, in the long term, develop ailments as a result of the food they eat, the work they do and the conditions they are kept in. One of the main causes of rheumatism is the build-up of toxins in the connective tissue. This is aggravated by diet, stress, and the way in which the horse is worked. This process can take years to build up and in most cases it will go unnoticed. It is only in later life that conditions such as rheumatism will appear. Bear this in mind when treating the horse: it took many years for this condition to develop, so do not expect to cure it in a matter of days or even weeks.

Firstly attention must be paid to the diet. Avoid foods which are highly processed, contain artificial additives and preservatives along with refined sugars. Feed a simpler diet with the addition of plenty of green and root vegetables, rich in vitamin C. **Cider vinegar** can be added either to the feed or the drinking water at the rate of between 30-50ml per day. This diet, along with the correct herbs, will help the horse's system eliminate toxins and waste products.

Obviously, even if the diet is changed, an improvement will not be achieved if the other contributory factors are not addressed. Identify the cause and remove it from the picture, whether it be stress, methods of training or living conditions.

With regard to your choice of herbs there are a huge number of what could loosely be called 'antirheumatic' herbs. These herbs have a variety of actions on the body – e.g. alterative, anti-inflammatory, circulatory, stimulatory, diuretic, analgesic, and digestive – and should be chosen depending on where you think the horse needs help. *It is not necessary to give them all.* Improve the diet if necessary, identify where you feel the horse needs help and then choose the appropriate herbs.

ALTERATIVE HERBS: now more commonly known as blood purifiers, will gently cleanse and restore the proper function of the horse's body. These include herbs such as **celery seed, clivers, burdock, kelp** and **nettle.**

ANTI-INFLAMMATORIES: which are used to support the body in its natural action to reduce inflammation. Herbs such as **meadowsweet, chamomile, devil's claw** and **calendula**. Do not use these in isolation; recognise that inflammation is an indication of something wrong, such as damage, infection or toxic build-up.

CIRCULATORY HERBS: will cleanse the toxins by increasing blood flow to the affected muscles. These could include **rosemary** (I have found rosemary to be particularly effective with rheumatic conditions), **buckwheat, comfrey, nettle** and **hawthorn.**

DIURETIC HERBS: will support the kidneys in their action to eliminate body waste, toxins and the waste products of inflammation. Try **dandelion leaf, meadowsweet, nettle, yarrow, couch grass** or **celery seed**, whose volatile oil seems to be particularly good for rheumatic muscles.

HEPATIC HERBS: support the liver in its work of cleaning and eliminating the waste and toxins which cause the inflammation. These include herbs such as **dandelion root, meadowsweet** and **milk thistle.**

LYMPHATIC HERBS: will help the action of the lymphatic system in its role of tissue cleansing – **calendula, clivers, fenugreek.**

Externally, herbs or oils can be used on the affected muscles to stimulate the circulation in the area, thereby increasing blood supply, which will relieve inflammation.

On areas where there is pain **hypericum** oil can be massaged in gently. Essential oils such as **lavender, chamomile** and **rosemary**, correctly diluted with a carrier oil, can be massaged directly onto the affected area.

A warm compress of **rosemary** or **thyme** can be used – make an infusion with 30 grams and 1/2 litre/1 pint of boiling water. Soak the compress in the infusion and apply (taking care that it is not too hot), bandage in place and repeat every 4 hours. If the skin is not broken in the area then **arnica** tincture is excellent for rheumatic, sore, bruised or inflamed muscles: add 5 ml/1 teaspoonful of tincture to 1/2 litre/1 pint of hot water, soak the compress and apply as warm as possible (taking care that it is not too hot), bandage in place and repeat 2-3 times a day if possible. In the past I have made a lotion using

5ml **arnica** tincture, 5ml **comfrey** extract and 5ml **lavender** oil mixed with 500ml aqueous cream and massaged it into affected muscles. This is a particularly good lotion to use if the area is difficult to compress. The **arnica** and **comfrey** will help with the inflammation, whilst the **lavender** oil will stimulate the circulation and help disperse the toxins which build up in rheumatic muscles.

I have known many horse owners who have successfully used copper pastern bands on horses with rheumatism. (*See* Useful Addresses section at the back of the book.)

RIGS AND FALSE RIGS

In very simple terms a **rig** is a male horse whose testes, either one or both, have failed for whatever reason to descend into the scrotum. They may or may not produce sperm and they may or may not be fertile. All, however, will produce hormones and behave like stallions.

A **false rig** is a male horse that has been completely castrated, but still displays stallion-like behaviour. This may vary from rounding up mares, to having an erection and attempting to cover.

Ownership of these horses can prove very difficult, especially if you are not in the position of owning your own land or being able to rent a secure paddock which allows you to separate the horse from mares. In mixed groups these horses can do a great deal of damage, ranging from harassing and attempting to cover mares to chivvying and attacking other geldings.

Many owners of false rigs find themselves in the position of having to keep the horse stabled at all times or run the risk of putting them in fields adjacent to mares. This can result in the horse trying to jump the fence to reach the mares or, as in one case I heard of, attempting to cover the mare over a barbed-wire fence, which resulted in extensive damage to both the horses involved.

I was told of this unfortunate event by the gelding's owner, who was desperate to try and solve the problem. After this incident she had been told that unless she could resolve the situation, she would have to move the horse because of the threat he posed to others. The owner contacted her vet, who was unfortunately unable to help her (the horse had been correctly gelded as a youngster).

The gelding's owner decided to give the horse 15 grams of **agnus castus** per day. Agnus castus has been used for centuries by both men and women, and was employed extensively in medieval times as an **an**aphrodisiac (to reduce libido) by religious novices and monks, who led a life of celibacy. (Hence its common name of Monk's Pepper.)

Within a fortnight his behaviour had altered dramatically, he had lost interest in the mares and was able to be turned out near them with no further problems. No longer worried and stressed, his weight increased and his mental attitude improved beyond recognition. The horse's owner reduced the agnus castus to 5 grams per day after a month, and a few weeks later the herbs were dropped altogether, with no recurrence of the 'riggy' behaviour.

NOTE: It must be remembered that when horses are gelded, especially later in life, the hormones present in the body at the time of gelding will not disperse immediately. Patience must be shown to these horses, who may still display 'full' horse characteristics for several weeks or months. The horse must be allowed time to adjust to his new condition.

RINGWORM

Despite its name, ringworm is actually a fungal infection. The fungi are able to survive for over a year in crevices and cracks in stables, farm buildings, horse boxes and in wooden fencing. However, it is far more usual for the horse to become infected as a result of a new horse suffering from ringworm, but not yet showing the visible signs, being brought into the yard or by coming into contact with an infected horse at a show. It can also be passed on by using infected rugs or numnahs (saddle blankets), or by being groomed with infected brushes. As soon as the condition has been identified, the horse must be isolated from other horses on the premises and its tack, rugs, grooming kit etc. disinfected. Everything that is used on the horse must be kept scrupulously clean.

NB: Remember to take care when handling horses with ringworm, as this condition can be contracted by humans.

The condition must be treated both internally and externally.

INTERNALLY: You must use herbs that will strengthen the body's resistance to infection and support lymphatic drainage. Herbs such as **echinacea, garlic, clivers, calendula,** and **cider vinegar.**

EXTERNALLY: Use antifungal herbs such as **calendula** and **echinacea** in tincture or lotion form. **Cider vinegar** can be effective if rubbed directly onto the lesions at least four times a day. Oils (correctly diluted) with anti-fungal actions, such as **calendula, lavender,** and **tea tree,** can be used. An old remedy for ringworm was to make a brew using lemon juice, crushed cloves of **garlic** and **cider vinegar.** This was painted onto the affected areas two or three times a day, taking care to go well beyond the fungal ring. To help encourage hair regrowth after the infection has cleared, a stimulating rub containing **rosemary** oil is recommended. A brew of **burdock** root or an application of **castor oil** are also traditional remedies for hair growth.

SARCOIDS

ANGLEBERRIES

These are a common type of semi-malignant skin tumour thought to be caused by a virus. They can have several different appearances. As far as herbal treatment is concerned it is not necessary to differentiate between the types. They can appear anywhere on the body, but in general tend to be found on the sheath, between the front legs, on the thighs and girth line or on the head. There are now several different methods of dealing with them via conventional medical treatment, including excision, cryosurgery, BCG injections, laser surgery, and more recently a new cream and injection produced by the Liverpool Veterinary University which has been successful. My feeling is that before resorting to some of these quite drastic treatments it is worth trying the non-invasive use of herbs.

Sarcoids should be treated in much the same way as we approach any other invasion of the body, by strengthening the body's own resistance and encouraging it to reject the virus. I have seen this rejection and it can sometimes occur within a few days of giving the herbs. The sarcoid appears to be pushed out of the body, where it then dries and withers, only to drop off shortly afterwards. This is not always the case; however, as a general rule of thumb, if the herbs are going to be effective they will usually produce some sort of changes within 6-8 weeks. Do not be alarmed if the sarcoid appears to 'explode' – this is a good sign. The area may become rather messy at this time and care must be taken to keep the site clean and free from flies. I would certainly recommend that, if possible, the herbs be administered during the winter months, when there is less likelihood of the horse being bothered by flies and heat.

Use herbs that have a cleansing, antimicrobial, and 'anti-tumour' action, as listed below.

Garlic – antimicrobial and anti-tumour. Recent trials in China have shown that garlic is effective in reducing tumours. Old remedies for tumours, which date back as far as the ancient Egyptians, advocate making a paste of fresh garlic and castor oil and rubbing it into the tumour.

Fenugreek – an excellent sister herb to garlic; it has been used in China for centuries in the treatment of cancer and tumours.

Rosehips – rich in vitamin C and volatile oils to help fight infection and strengthen the body's resistance to invasion.

Clivers – this herb has been shown to boost the production of white blood cells. Use in conjunction with **calendula** for the lymphatic system, to help cleanse the body of toxins and waste products.

Echinacea – antimicrobial and immuno-stimulatory.

Burdock – has been used to treat tumours and has been shown to inhibit tumour growth.

Wormwood – one of the old Plague herbs, a bitter and traditionally used for tumours.

Meadowsweet – used for its anti-inflammatory effect to help once the sarcoid has started to respond to the herbs.

Whilst the sarcoid is changing it can be additionally helpful to use a cream on the area. Ointments containing **calendula, garlic** and **vitamin E** are particularly effective and can be used on the site after the sarcoid has dropped off, to encourage the skin underneath it to heal quickly.

An old horse of mine that suffered from sarcoids, showed an insatiable appetite for common thistle heads, wild hops, and vine leaves and tendrils. I let him eat them whenever we encountered them and in addition gave him some of the herbs recommended above. His sarcoids were quickly shed and never reappeared.

There are several homoeopathic remedies available for both sarcoids and warts. These include **thuja,** which can be given in tablet form, as well as in a cream or tincture for external application. Speak to a homoeopathic vet for details on dosage and other remedies available for this condition. I have found in my experience that horses seem to respond to one method of treatment or the other. Those horses who do not respond to the herbs have had success with the homoeopathy and vice versa. If necessary your homoeopathic vet can make a 'nosode' if he/she is supplied with a little sample of the sarcoid. This is like producing a homoeopathic 'vaccination'.

SCOURING/DIARRHOEA

Scouring or diarrhoea may have a number of causes. These could include feeding an over-rich diet; bacterial or viral infections of the intestines; stress; over-excitement caused by competing or travelling, resulting in disruption of the gut bacteria; sudden change of diet; or even something as simple as a change of routine. In most cases it is the horse's way of cleansing digestive poisons from its system and in these instances the problem can usually be resolved within a day by using the appropriate herbs, a sensible diet and, if necessary, **probiotics.** If, however, the horse shows no sign of improvement within 24 hours then it would be advisable to ask your vet to have a look at the horse.
(*See also* Dehydration.)

There are, of course, more serious causes, such as heavy parasitic infestation, ulceration and stomach tumours, to mention but a few. If any of these are suspected then a vet should be called in immediately.

It is worth mentioning at this point that many horses will have a tendency to produce soft, loose droppings when first turned out onto fresh spring grass. Additionally it is not uncommon for a certain amount of liquid to be excreted with the droppings. This is nothing to worry about and is quite normal.

Oak is high in tannin and as such is both tonic and astringent. My horse, who has an oak tree in his field, always eats a certain amount of the young leaves in springtime to counteract the loosening effect of the spring grass. Once the summer months approach he shows no interest in the tree at all. (NOTE: Horses should **not** be allowed to eat acorns.)

Another mild digestive astringent can be found in **meadowsweet**, and **comfrey**, with its demulcent, soothing and healing action, is ideal if inflammation of the gut is suspected. **Slippery elm** bark is specifically for digestive problems and diarrhoea. It is nutritious and will act as an internal poultice by soothing and relieving inflammation and irritation. I find that it is best to mix the slippery elm powder with a natural live yoghurt and add this to the feed or syringe it directly into the horse's mouth. The recipe is 1 tablespoon of slippery elm mixed with 2 tablespoons of natural live yoghurt, and given at least twice a day. I use this in preference to the pharmaceutical products available; these tend to be chalky compounds which 'bung' the horse up, and the end result being, that once discontinued, the problem often re-occurs.

Live plain **yoghurt** is very helpful in cases of scouring as its natural bacterial content will help re-balance the gut bacteria. This can be fed with honey, either in the feed or mixed to a paste and syringed into the horse's mouth. Give 2-3 tablespoons of yoghurt mixed with 1 tablespoon of honey three times a day.

Probiotics have a big part to play in cases of diarrhoea or scouring and can be used to great effect in returning the gut bacteria to normal. They are best obtained from your veterinary surgeon or from a reputable manufacturer. (For more information on probiotics see Non-Herbal Miscellany section; for suppliers see addresses section at the back of the book.) Reputable manufacturers of these products and veterinary surgeons will often carry out faecal analysis for horse owners to help try and ascertain the cause of the problem.

SHEATH CLEANING

In general this is a task that can be carried out very simply and easily by using a small sponge and a bucket of warm water. Avoid using strong disinfectants or soaps as these may kill the bacteria that live on the penis, thereby encouraging the development of infection. Most geldings and stallions, if handled correctly from an early age, will happily allow their sheaths to be cleaned and the build-up of the debris, dirt, mud and smegma carefully removed.

Some horses, however, will not allow you near this area and in these cases debris and smegma can build up and harden, which in turn can cause irritation to the penis inside the sheath. On these occasions I have found a remedy used by an American herbalist to be very helpful. She told me how she uses **uva-ursi** to help geldings and stallions whose sheaths have become infected, are difficult to clean or have an over-production of smegma. Since then I have used this treatment successfully on my stallion and a couple of geldings. I must stress, though, that this is not a short cut to good horse management, and this approach should only be used when the horse will not allow normal sheath cleaning or if infection is present, and only for a short period. Give 1-2 leaves daily to ponies and 3-4 leaves daily to large horses for seven days only.

STALLIONS AT STUD

During the stud season it is essential to wash off the stallion's penis immediately after covering. For washing my own stallion I use 5 drops each of **hypericum** and **calendula** extract for their cleansing, soothing and antiseptic actions, in approximately 10 litres/2 gallons of warm water. There is no reason why this solution should not also be used for normal sheath-cleaning purposes.

SHOCK

THE FOLLOWING INFORMATION IS APPLICABLE TO BOTH HORSE AND RIDER.

For horses that have had a traumatic experience, such as a bad accident, fall, loss of a close companion or change of home, give the **Bach Flower Rescue Remedy** immediately. Either deposit 4-6 drops straight into the mouth, or if the horse will not allow this, put 4-6 drops onto a piece of sugar, apple, etc. and feed. Some horses will lick it off the back of your hand. Put 10 drops into the horse's bucket of drinking water for several days after the shock. **Aconite** and/or **arnica** in homoeopathic form, if given shortly after the event, are both excellent remedies for shock. If you have prior warning of the impending trauma, for example if you know the horse is moving or losing its friend, give the Rescue Remedy for several days prior to and during the stress period. There are specific Bach Flower Remedies which will help in coping with the upheaval of moving stable or home – see the section on Bach Flower Remedies in Part Three for full information.

Skin Condition

See Coat and Skin Condition.

Skin Irritation

See Coat and Skin Condition, Dermatitis, Eczema, Mud Fever, Sweet Itch.

Soft Swellings or Oedemas

These can be caused by, amongst other things, a blow to soft tissue, lymphangitis, enlargement of the tendon sheath (windgalls or windpuffs) or as an allergic reaction to food, insect bites or vaccines.

Whatever the cause, the end result is usually inflammation and bruising, with an increase in fluid caused by a build-up of waste products. The swellings, or oedemas, can often disperse of their own accord within a few days, once the inflammation has subsided and the lymphatic system has done its work in clearing and cleansing the area of waste products. Some swellings, however, can prove stubborn, and although they do not often appear to be painful or cause lameness, they are unsightly, particularly if you wish to show the horse.

Personally I am strongly opposed to draining these soft swellings or injecting cortisone, which is often recommended in these cases. Nature fills a vacuum, and draining the area will usually result in the fluid returning within a few days. It is far better to try and support the horse's system in dealing with any inflammation present and to stimulate the circulatory and lymphatic systems to encourage the fluid to disperse naturally. This can be done by using herbs, both internally and externally.

Obviously if the swelling has been caused by an allergic reaction to food, or insect bites, then the horse must be removed from the cause of irritation, either by changing the diet or by protecting the horse from the insects. Use herbs internally that will tone and support the lymphatic system, such as **clivers** and **calendula**. **Comfrey** is ideal in these situations: it will encourage healing, reduce the bruising and soothe the inflammation. Circulatory stimulants such as **nettle** (make sure the horse does not have an allergy to nettle!), **buckwheat** and **hawthorn** will also help to cleanse toxins by increasing the blood flow to the affected area. Diuretic herbs such as **dandelion, celery seed** and **couch grass** can help remove excess fluid in the body. If inflammation is present then use **meadowsweet** or **devil's claw**. I always tend to add **cider vinegar** to the horse's water or feed at this time as I have found it helpful in encouraging the absorption and dispersal of fluid.

Externally, the use of a compress or poultice can be helpful, although many soft swellings tend to occur on areas of the body which are impossible to bandage! Once again **comfrey** is a specific for any sort of soft swelling. Make a comfrey poultice or compress, or use a pulp of fresh comfrey leaves and bandage onto the area. If this is not possible, comfrey oil or ointment can be gently massaged in to help reduce the swelling. Other compresses that can be used are **chamomile** and **calendula.**

SWEET ITCH

SUMMER ECZEMA, SUMMER DERMATITIS

This infuriating condition, which causes intense skin irritation, is the result of an allergic reaction to the bites of a particular midge – Culicoides.

It is interesting that this allergy, which we usually associate with ponies, is now being reported amongst such horses as the European Warmbloods. I feel sure that this is due to the fact that more of these horses are being exported from their native lands to Britain and the US, where they may be exposed to the midge for the first time. In Britain we see a similar situation in our native ponies, who whilst living in the wild do not suffer from this problem, but develop it once taken out of their natural environment and kept in less harsh conditions. Another example of how change of location can affect horses can be seen in the instance of the Icelandic ponies. In their native country, where the air is particularly clean and free from pollution and humidity, they have no sweet itch problems. However, it has been found that in many instances these same ponies will develop the condition within two years of being exported to Britain and elsewhere.

The important thing to remember when dealing with sweet itch herbally is that if you know your horse is prone to sweet itch then you must, if possible, try to use herbs prophylactically rather than curatively.

Take whatever steps you can to remove the horse from the midges by, if possible, keeping it in during the worst periods, i.e. daybreak, early evening and during warm humid days. Try to use an effective fly repellant, and if the horse cannot be stabled at the critical times, provide some sort of cool, shady area or field shelter.

Internally, prior to the onset of the season, give the horse or pony herbs that will strengthen the defence systems, and help cleanse the blood toxins. Sulphur is important as a blood purifier and is essential for healthy skin and hair. Herbs such as **garlic, calendula, nettle, kelp** and **meadowsweet** all contain sulphur and can be successful. **Clivers** will support lymphatic cleansing, as well as strengthen hair growth. Herbs that have an antimicrobial action, in that they stimulate the production of white cells, such as **echinacea, garlic** and **calendula,** can be given to horses either before the onset of sweet itch or to those already suffering from the condition. It is uncertain whether antihistamines have an action on horses, but it may be worth considering **buckwheat** if the horse already has the problem, as it is said to have an antihistamine action.

Cider vinegar should be given internally, 30ml a day, and an effective rinse can be made by adding 2 tablespoons of cider vinegar to 1 litre/2 pints of water and rinsing the mane and tail. The vinegar's astringent and cooling action will reduce irritation and act as a repellant to midges. If it is already too late and the horse has rubbed itself, creating lesions and that characteristic 'elephant skin', then do not use the cider vinegar rinse as this may sting, but do give the herbs internally and apply an ointment containing any of the following: **calendula tincture, garlic, comfrey, hypericum,** or **propolis.** The **Bach Flower Rescue Remedy** cream can be useful in some instances, especially where there is considerable trauma to the skin.

Homoeopathic remedies can offer relief to horses and ponies suffering from sweet itch. One remedy, called **Culicoides midge**, is made from the midge itself, and if given during the season can be helpful. There are a number of other remedies which may be applicable depending on the exact nature of the sweet itch, such as **arsenicum album**, and **sulphur.** However, when using homoeopathic remedies it is best to consult a homoeopathic vet, who will be able to advise you on which remedy is most suited to your horse's individual case.

Tendon & Ligament Strain/Damage

Although tendons and ligaments are two separate structures, with different functions in the body, they are both treated with the same range of herbs when they have suffered injury or damage. With both tendons and ligaments, damage or strain can mean long-term rest to allow the structures to repair. Often the horse will appear to be sound, but if brought back into full work too soon, a further breakdown will result.

In most cases immediately after the damage has been done, be the cause a blow, rupture, strike or strain, there will be inflammation, swelling, bruising and pain. These symptoms can disguise the true extent of the damage and it is vital that a vet be called to assess the situation.

Most horses will be prescribed box rest, at least until the inflammatory swelling has subsided, along with bandaging of the damaged area. (Remember to bandage the other front or back leg, as in attempting to rest the damaged leg, the horse will put additional strain on the tendons of the undamaged one). Cold compresses can be made with either a **comfrey** infusion or a rinse made with **arnica** and **ruta grav.** tincture. To make the rinse, add 1 teaspoon of each tincture to 1 litre/2 pints of water. Do not use this rinse if the skin is broken as the tinctures could cause irritation. Both the comfrey and arnica compresses will help reduce any bruising and inflammation present, whilst at the same time encouraging the circulation. Change or resoak the compress with the chosen infusion as often as necessary. Remember: for horses with sensitive skin, watch for signs of any reaction to the compresses.

Internally, herbs that will reduce inflammation and pain should be given, along with herbs that will stimulate the circulation, encouraging the removal of inflammatory fluid. Use herbs such as **meadowsweet** or **devil's claw**, for their anti-inflammatory and analgesic action; **comfrey,** to reduce bruising and encourage healing; **buckwheat** or **hawthorn,** to stimulate circulation; and **clivers, calendula** and **kelp,** to support the lymphatic and glandular system. **Cider vinegar** can be used both internally and externally for its astringent and blood-cleansing properties.

For horses that are restricted to long periods of box or stall rest due to severe tendon or ligament damage, see the section on box rest. Remember: recovery can be a long, slow process and patience is the watch word.

Homoeopathic remedies can prove very effective for tendon and ligament problems, both immediately after the injury and for helping with renewal and recuperation. Contact your homoeopathic veterinary surgeon for advice.

THRUSH

This condition occurs when the frogs of the feet become infected. The clefts either side of the frog accumulate a foul-smelling material. Depending on the severity of the infection, the horse can become lame as a result. If the condition goes untreated the infection will spread to the sensitive tissues.

Thrush occurs when the feet, and especially the clefts, are not picked out correctly and regularly. Too much time spent standing on damp dirty beds will encourage the condition, and horses with deep clefts are especially at risk. Firstly ensure that the horse's bed is absolutely clean and dry, then call the farrier and ask him to trim the frogs, removing all dead tissue. Scrub the soles and clefts with a strong salt solution. Make a dilution of either **tea tree** or **lavender** oil by adding 5 drops of the oil to 30ml of a suitable carrier oil (grape seed, sunflower, walnut), and paint the whole area two or three times a day. Scrub the feet with the salt solution before each application. Feed the horse with herbs such as **garlic, echinacea, rosemary, rosehip,** and **kelp.**

These will help the horse to fight infection and encourage strong healthy hooves and hard soles.

If the infection is already too advanced, you may need to call your vet, who will probably advise a course of antibiotics and the use of an antibiotic spray on the affected area.

ULCERS – MOUTH AND GASTRIC

For eye ulceration, *see* Eye Problems.

MOUTH ULCERS: Mouth ulcers tend to be an indication of a poor state of health or depressed immune system. They may occur in horses after periods of ill health, viral infections, after a course of antibiotics or after having come into contact with chemical sprays or toxic plants. Normally a blood test is taken to try and ascertain the state of the horse's health, but more often than not nothing conclusive is found. One woman told me she eventually discovered that her horse's ulcers where caused by a fungal infection he picked up from his hay.

Whatever the cause the overall health of the horse can be improved by feeding a good cleansing diet which includes plenty of fresh fruit and root vegetable. **Oats** and **garlic**, being rich in B vitamins, are excellent to strengthen the nervous system; **rosehip** and **nettles** will provide plenty of vitamin C. As a good all-round vitamin and mineral supplement, **kelp** is hard to beat; and **comfrey,** which is rich in B12, will encourage healing.

A mouth wash can be made using **sage**. Either make an infusion and syringe into the mouth, or give the horse a few fresh leaves to chew on – but remember: you should not use sage on a pregnant mare.

NB: Remember also that poorly fitting or sharp bits can cause mouth ulceration.

GASTRIC ULCERS: Gastric ulceration is a complaint we are starting to hear more about. It must be taken as an indication that there is something very wrong with either the diet of the horse or its state of health and mind, or both. A recent study was carried out in the US and Hong Kong on Thoroughbred racehorses[1] and an abnormally high incidence of ulceration was reported. These alarming results are being attributed to a combination of causes, including high stress levels, incorrect feeding, and the over-use of drugs.

As with all herbal treatments the horse must be viewed as a whole, and although herbs can be given that will produce good and fast results, you must not lose sight of the overall picture. If the conditions that first caused the ulceration are not dealt with, they will quickly return. Therefore look closely at the horse's diet to ensure that this is not contributing to the problem. If stress is a key factor then action must be taken to identify the cause and remove it. Acidity in the gut can aggravate ulcers and the use of **probiotics, natural live yoghurt** and **honey** can help to counteract the effect of excess acidity. (See Non-Herbal Miscellany for details of probiotics, honey and yoghurt.) Certain drugs can, at worst, cause ulceration or at least aggravate the delicate mucous membrane of the gut wall. Drugs place additional strain on the horse's system and can result in horses who are not in good health being competed. This situation must ultimately result in the horse being put under even more stress.

The following herbs have produced excellent results when used on horses with ulceration of the gut:
Slippery elm – this will act as an internal poultice, reducing over-acid secretions and soothing inflammation.
Comfrey leaf – this is a specific for gastric ulcers; its mucilagenous content will soothe and heal.
Marshmallow root – being demulcent and mucilagenous, it is ideal for the treatment of inflammation and ulceration of the digestive system.
Liquorice – excellent for ulcers; when ingested, it produces a thick sticky mucus which will reduce gastric acid and encourage healing.
Meadowsweet – particularly effective when used for ulcers resulting from drug use or poisoning. It is one of the best all-round herbs for the digestive system and will soothe mucous membranes in the digestive tract, reducing excess acidity and associated pain.
Valerian and chamomile – ulcers, be they in humans or animals, are often caused and certainly aggravated by stressful conditions. The way a horse is kept, what it is fed, how it is worked and the competitive job it is expected to do, can all create stress, agitation and the possibility of resultant ulcers in exactly the same way as they can be created in humans who live on a poor diet, smoke and drink too much or work in high pressure careers. If stress is indicated in the condition valerian and chamomile are the best herbs to use in conjunction with one or more of the other herbs listed here. By helping to calm and soothe the horse's nervous system the resulting stress can be reduced and the horse can be given the opportunity to start healing itself.
Milk thistle – although having a demulcent action, milk thistle is not specific for ulceration. It is used for its overall beneficial effect on the liver. This organ's role is to cleanse the blood of toxins and waste materials, and an incorrect diet and the use of drugs will produce additional toxins that the liver has to deal with. Use of milk thistle will protect and support the liver function, which must ultimately lead to an improvement in the overall health of the horse. This must be the goal to work towards in cases of gastric ulcers.

WARTS

Warts and their various cures are steeped in folklore and tradition. The list of old remedies is too lengthy to mention, and they were not always very successful anyway. In country districts of Britain the 'wart charmer' was a familiar member of the community, as was the local blacksmith, whose water tub used for cooling hot iron was much sought after as a wash for warts. One of the oldest remedies and one that still produces results these days, is **dandelion,** whose milky sap rubbed onto warts can effect a cure.

Warts are a viral infection which can be transmitted in a number of ways, e.g. foals can receive the virus from their mothers whilst suckling. In general, they seem to affect younger horses more than adults.

There would appear to be two fairly distinctive types of wart. Firstly, there are 'milk', 'grass' or 'deciduous' warts, so named because they tend to disappear or drop off spontaneously within a few months. Grass warts are normally small and can cover large areas of the muzzle, nostrils, eyelids and front legs. They tend to be found on young horses and do not usually seem to bother the animal, although they are unsightly, and if knocked can bleed or become infected. The second type of wart tends to be much larger and may appear on any part of the body, but mainly on the face, legs, sheath or ears, and will not normally disappear of its own accord. In both cases the aim must be to eliminate the viral infection from the horse, rather than resort to the drastic methods of cutting or burning them out which are favoured by conventional veterinary medicine. These methods, although much quicker, may cosmetically improve the horse's appearance but will not even begin to tackle the inherent problem.

As with any viral infection the first action must be to strengthen the horse's resistance to the virus and encourage the rejection of the wart. I have witnessed warts being shed within the space of one week following either herbal or homoeopathic treatment, although one should not count on such a dramatic reaction, and sometimes several weeks will elapse before the wart is shed.

Use herbs that will cleanse, strengthen and tone the body. Give herbs rich in vitamin C, such as **nettle** and **rosehip,** along with lymphatic cleansers and tonics, such as **clivers, calendula, garlic,** and **wormwood. Echinacea,** with its antimicrobial and immuno-stimulatory action, will strengthen the horse's resistance to viral infections, and can be used along with **burdock** and **dandelion** for the hepatic system.

The homoeopathic remedy for warts is **thuja**, which can be used both internally in remedy form and externally as an ointment. The ointment should be applied to the wart as often as possible, preferably 3-4 times daily. Thuja tincture is also available and this can be painted onto the wart 2-3 times daily. **Do not use the thuja plant internally**.

Greater celandine can be used as an external application, but **do not use greater celandine internally.** Squeeze the juice out of the fresh stems and apply directly onto the wart. After the wart has been shed, encourage the healing of the skin with an ointment containing **vitamin E** and **calendula.**

WEAVING

Weaving is generally regarded as a stress-related problem comparable with the 'behavioural ticks' often seen in caged and zoo animals. There are some instances of foals picking up the habit from their mothers if they spend a lot of time in the stable together, but most horses tend to develop the habit as a result of too much time spent confined to stables and as a response to stress situations such as feeding times and the removal of stable mates etc.

As with any stress-related condition the ideal action is to remove the cause of the stress, and certainly if horses can be allowed out to graze as much as possible this can make a big difference. However, I have seen one extreme case where the horse in question continued to weave standing in the field. Anti-weaving grilles can be fitted to the stable door, which may resolve the problem, but some horses just stand a little further back from the door and continue to weave.

Long-term, the worst physical effect is on the horse's tendons, which are put under enormous strain from the continuous rocking motion. In addition, the horse can suffer weight loss and even recurrent spasmodic stress-related colic as a result of the anxiety he is feeling. (For advice on herbs for stress in the gut see section on Spasmodic Colic (under Colic) or Nervous Gut.)

If you can, try to keep the horse out of the stable and occupied as much as possible. It is no coincidence that many of the worst weavers are quick-witted, busy, intelligent types that respond well to work and a varied routine. If it is the feeding routine that is the problem, try and break it if possible – for example, feed the horse out of the stable or in the field. Wherever possible try and remove the stress factor from the horse. Herbs such as **chamomile** and **valerian**, which will calm and relax the horse, are ideal in that they will soothe the nervous system without doping the horse.

I am pleased to say that aromatherapy is now gaining acceptance amongst the horse fraternity, and in these situations it comes into its own. Use the old stallion man's trick of putting a few drops of an essential oil such as **lavender** or **chamomile** onto a piece of cotton and attach it to the horse's headcollar on or near the noseband. The soothing effect of these oils can make a big difference to the mental state of a horse, helping it to remain calm in stressful situations.

WORMS

The horse can play host to a wide variety of parasites including: bots, roundworms, pinworms, threadworms, large and small strongyles, tapeworms and lungworms. These worms can cause untold damage, especially those that migrate around the system, and if horses are not protected from an early age worm infestation can result in, amongst other things, recurring colic attacks, intestinal bleeding, lung damage, scouring, liver damage and death.

I am not going to suggest that herbs are the whole answer to a complete worming programme. However, they have an extensive part to play in keeping the horse healthy, and resistant to infestation and can be used safely alongside pharmaceutical wormers. Couple this with good horse and paddock management and you can protect your horse from these dangerous parasites. It is unfortunate that our horses' paddocks are now, in the main, devoid of Nature's worm expellants, such as mustard, couch grass and brambles.

French folklore states that herbs with worm expellant properties should be given at the waxing – near the full of the moon – when the worms are migrating and are easier to dislodge. The herbalist monks of medieval England advised the use of **garlic** when treating worm infestations in animals. **Wormwood**, as its name suggests, has strong anthelmintic properties, but should be used with care. Give foods that the worms do not like, such as **cider vinegar, garlic, grated carrots**, and seeds such as **pomegranate, melon, pumpkin, mustard, fennel** and **aniseed** (about twice a month feed the large seeds at the rate of 15 grams daily, or the small seeds at around 10 grams daily). Let the horse graze on **brambles** whenever possible and give **couch grass**. Strengthen the horse's overall health by using herbs that will support the liver such as **dandelion, burdock** and **milk thistle**. All these herbs, foods etc. can be combined on a regular basis with the horse's normal diet.

In addition to this herbal regime take as much care as possible with the horse's pasture management. It is a time-consuming task, but if grazing is restricted then the droppings must be picked up daily to break the parasitic cycle. If possible, graze cattle or sheep with the horses, or run them onto the land after the horses have been moved.

If the horse has already suffered worm damage and the intestines and liver have been compromised, then immediately instate the regime covered above and use herbs that will support liver regeneration and heal the gut. **Milk thistle** and **liquorice** are specifics for encouraging liver cell production and improving liver function. **Comfrey, meadowsweet,** and **marshmallow** will reduce inflammation and irritation and encourage healing, whilst **kelp** will detoxify the system and strengthen the horse's resistance.

Homoeopathically, there are remedies that can be given either in a preventative way, or to encourage healing if worm damage has already occurred. Contact your homoeopathic veterinary surgeon for further details.

WOUNDS

Wounds come in all shapes and sizes and if there is any question as to their severity then the vet should be called immediately. Fortunately, however, most wounds that horse owners experience tend to be fairly minor and normally just involve cleaning and protecting from infection. Vets will tell you that if the wound is serious enough to require their attention, they would prefer that it is not cleaned out with disinfectants and then filled with various ointments. It tends to make their work harder if, before they can examine the wound, they have first to remove a plug of sticky ointment. If a wound does require veterinary treatment then simply wash out the site using a solution of warm water with a few drops of **hypericum** tincture added. If hypericum is not available then use common salt dissolved in water.

If, however, you are just dealing with a small graze or cut, then once again wash the wound as described above, ensuring that all grit and debris are removed. If the wound is still bleeding, gently bathe with **distilled witch hazel**, which has an astringent and blood-clotting action. If this is not available then make a compress from any of the other astringent herbs available, such as **hypericum, golden rod, rosemary** and **yarrow**, all of which will speed up blood-clotting. If the wound has stopped bleeding then an ointment made from **comfrey** is the best and most effective healer available. Comfrey contains allantoin, which stimulates cell division and will speed up the formation of scar tissue. Do not, however, use comfrey if the cut is very deep as it is so effective a healer that the outer surface of the wound may heal before the inside does. For deep wounds use hypericum compresses. There are now many excellent all-purpose wound ointments available containing **hypericum** and/or **calendula.**

Internally give **garlic** to help the horse fight any infection that may arise as a result of the wound.

I find the **Bach Flower Rescue Remedy** invaluable when dealing with wounds and will give the horse 4 drops on a piece of apple or off the back of my hand as soon as I discover the wound. There is always the likelihood of stress with injury, which will slow down the recovery time.

Bruising, to a lesser or greater degree, frequently accompanies wounds, and the homoeopathic remedy **arnica** is always handy to have in the first aid kit. The remedy can be given immediately after the incident without having to worry about it interfering with any treatment the horse may require if the vet has to be called. Contact your homoeopathic vet for general instructions on giving arnica in these instances.

PUNCTURE WOUNDS: Puncture wounds can be extremely troublesome and often go undetected; this allows infection to build up without the owner being aware of any injury.

They can be caused by, amongst other things, thorns, nails, pieces of wire or splinters of wood. When the foreign object (e.g. a blackthorn or splinter) is driven in, it can become lodged deep in the wound along with bacteria. This can result in deep infection, whilst externally the wound remains virtually invisible.

Internally, give herbs such as **garlic, echinacea, calendula** and **cider vinegar,** all of which have antimicrobial action, to help the horse fight any infection.

With puncture wounds the two main aims are to draw out any foreign objects or pus that may be present, whilst at the same time keeping the wound open to allow the free drainage of any foreign object, pus and infection.

A **slippery elm** poultice is ideal for these situations. It has powerful drawing properties and can be used as a medium to which antimicrobial, antiseptic and antibacterial agents such as **chamomile, lavender, eucalyptus, tea tree** or **garlic oil** can be added. To make a poultice, mix 1 or 2 tablespoons of slippery elm with boiling water to form a soft paste. If necessary, add a few drops of your chosen oil to the paste and apply to the affected area (taking care to ensure that it is not too hot), then cover and bandage. Change the poultice regularly, at least three times a day. Poulticing should not be prolonged. Consult your vet.

If the wound is in the sole of the foot it can be helpful to 'tub' the horse's foot in a bucket of warm water containing 10 drops of one of these oils, and then poultice with slippery elm powder to which a few drops of your chosen oil has been added.

Part Three

Alternative Therapies

Alternative Therapies

Bach Flower Remedies

These remedies were formulated in the 1930s by Dr Edward Bach, a medical consultant, bacteriologist and homoeopath from London's famous Harley Street.

I have used a variety of these wonderful remedies over the past twelve years and have found them invaluable for myself, my animals and even the plants in my care. All of the remedies are prepared from wild flowers, shrubs and trees and are preserved in grape alcohol.

Rather than tackling any physical conditions directly, the Flower Remedies aim to deal with the person's or animal's state of mind, and in restoring harmony to the mind and body will encourage it to 'heal itself'.

Each remedy, and there are thirty-eight of them, covers a particular aspect of negative states of mind. The thirty-ninth remedy, Rescue Remedy, is a composite of five of the Flower Remedies namely: cherry plum, clematis, impatiens, rock rose, and star of Bethlehem. Rescue Remedy is the one I have found most useful as an all-purpose comforter in times of stress, anguish or anxiety. I have used it successfully on horses, dogs, sheep and lambs, cats and chickens, not forgetting the horse rider.

For horses 2 drops of each chosen individual remedy or 4 drops of the Rescue Remedy can be fed either on a sugar lump or piece of apple or dropped directly onto the back of your hand for the horse to lick off. Alternatively 10 drops can be added to the horse's bucket of water, replenishing each time the bucket is refilled. It is excellent for horses that have had a shock, accident, distressing experience, or have lost a companion or are under stress of any sort.

All of the remedies are absolutely safe, and can be given in conjunction with conventional medicine. They are not addictive and the horse does not develop a dependence.

I have used the rock rose remedy, which is suggested for 'suddenly alarmed, scared, panicky, or terror', with excellent results on horses that suffer from panic attacks and on one of our dogs that is terrified of thunder storms.

I used the walnut remedy – 'assists in adjustment to transition, change or new surroundings' – on all the animals and ourselves when we moved home three years ago.

I have used the oak – 'normally strong/courageous, but no longer able to struggle bravely against illness and/or adversity' – on a horse that had been in a great deal of pain for a very long period and was starting to give up. He is, I am pleased to say, fully recovered.

I could go on and on, extolling the virtues of this excellent range of remedies. However, the best thing the reader can do is to obtain a copy of the Bach Flower Remedies' leaflet, which gives a comprehensive explanation of the Bach system, lists in full the thirty-eight remedies plus the Rescue Remedy, and gives much greater detail on the methods of application for both humans and animals. These leaflets are available free from any good health shop or by contacting the Bach Flower customer enquiries office in London (*see* Useful Addresses, page 167).

Aromatherapy and Essential Oils

Aromatherapy is a twentieth-century term, coined in 1928 by the Frenchman René-Maurice Gattefossé, a chemist in a perfume factory, who sustained a serious burn to his hand in a laboratory explosion and immediately plunged the hand into the nearest available liquid, a container of pure lavender oil. It was a most fortunate action.

Amazed by the exceptionally fast, non-infected and scar-free healing of the burn, Gattefossé studied, developed and wrote about the therapeutic use of plant essences (essential oils), launching the renaissance of one of mankind's most ancient healing practices, and possibly the one most closely allied to the use of herbs for therapeutic purposes.

The two disciplines do differ in various basic ways, but they are deeply rooted in the same traditions and make use of many of the same plant species. Infused plant oils and essences have stood beside fresh or dried herbs, decoctions and tinctures in kitchens and 'dispensaries' certainly since ancient Egyptian times, almost 3000 years BC, to be used alone and in flavourings, ointments, pastes and unguents. Their use in perfumery is legendary.

The great Arab physician Abu Ali Ibn Sina (Avicenna, AD 980-1037) is credited with the discovery of the method of extracting essential oils by distillation. Whether or not this is so, by medieval times 'chymical oils' as they were known, were in general use as remedies for a multitude of problems, many of which would most certainly have applied to animals as well as humans. A number of the oils remained in the pharmacopoeia well into the present century, until they were gradually supplanted by synthetic drugs.

Modern research into essential oils has underlined their antiseptic and bactericidal properties, as well as their balancing effects upon both mind and body, emphasising their value in holistic therapies. They are used in massage oils, baths, compresses, inhalations, ointments and lotions, as are other plant materials – but it is the unique combinations of chemicals and the aromatic elements of essential oils, with the subtle effects that these can have upon an individual person or animal, which sets them apart from all other substances and, for this reason, individual preference is usually taken into consideration when deciding which oils to use. The experienced aromatherapist will select a group of oils appropriate for the circumstances and then take note of these preferences, which can and often do change as the treatment progresses. Horses will usually 'state' their preferences very clearly, turning away from or reaching towards the oil offered.

The practising aromatherapist will carry a substantial armoury of essential oils for use in widely differing circumstances. There are, however, some simple remedies, using just a few oils, for everyday use with horses. These have been described in the entries covering the conditions where they may be found useful.

Essential oils can be bought in most health-food shops and also through specialist suppliers (see addresses section at the back of the book), as can basic carrier oils such as walnut, sweet almond or sunflower. Do buy the best quality natural oils with no additives (avoid using the synthetic or nature identical products which have recently come onto the market) and store in a cool, dark place.

It must be emphasised here that essential oils are very strong substances, some of which can be dangerous if incorrectly used. They should not be given orally without professional advice. If in any doubt whatsoever, consult a trained aromatherapist in conjunction with your vet before embarking upon treatments.

Dilutions suitable for the usages found in this book are:

1. For a basic massage oil:
10-12 drops of essential oil to each 30ml of carrier oil. Measure the essential oil into a bottle, add the carrier oil and shake gently to mix.

2. For a hot or cold compress:
Sprinkle 2-4 drops of essential oil onto each 1/2 litre/1 pint of water used and give a quick whisk round. Don't use too much water or you will risk wasting precious oil. Fold a clean piece of material (linen, towelling etc.)

into at least four thicknesses, dip into the water, remove and squeeze gently to wring out excess liquid. Use as quickly as possible.

NB: **Competition warning**. Some essential oils may contain certain constituents which are regarded as prohibited substances by official governing bodies, e.g. camphor, menthol and thymol. These substances, if used on horses even externally, could inadvertently be licked off and ingested. Although not harmful, they may cause the horse to 'test positive' in a blood test under FEI or Jockey Club Rules. It is therefore safer to avoid using essential oils for at least 7 days prior to competing under rules.

Further reading on the fascinating subject of aromatherapy can be found in the Bibliography.

The following case history was recounted to me by a woman who has been using herbs and homoeopathy on her horse for many years, and who as a result has become an acquaintance of mine.

Helen owns a wonderful black Welsh Section D gelding who is seventeen years old and goes by the name of Cracker. He is a particularly sensitive horse and has a history of allergy-related problems. On this occasion he had gone down with laminitis in all four feet and there was some suggestion that it could have been caused by a severe allergic reaction. He was confined to his stable on the vet's orders, where he ultimately spent eighteen weeks before he could eventually be turned out.

Cracker's condition deteriorated dramatically during the first four weeks of his confinement: both pedal bones rotated and he developed gas lines and abscesses in his front feet. He was in a great deal of pain and on high levels of drugs. Cracker made a new friend at this time – Cobweb, a pony that he became exceptionally close to and who became a grooming partner. Helen felt that because Cracker was no longer able to go out and have the companionship of other horses he came to rely heavily on his contact with Cobweb.

Unfortunately, not long after this, and, as if to add to Cracker's misery, Cobweb was moved to another yard. This seemed to be the last straw for Cracker and he lost all interest in the world around him.

During his worst period he was being given four sachets of phenylbutazone and eight acepromazine tablets a day. His liver and kidneys had been damaged, he was in terrible pain and mentally the horse was ready to give up. On several occasions Helen was tempted to put the horse out of his misery – he couldn't lie down, and, had he done so he probably wouldn't have got up again.

Helen rang me and said that the horse had lost the will to go on; he was incredibly depressed. He could not be encouraged to move; in fact he could not be persuaded to leave his stable, and she felt that it was now his mental, rather than his physical state that was preventing him from making a full recovery. I suggested trying some of the Bach Flower Remedies as well as contacting an aromatherapist, because it appeared to me that the only way the horse could be persuaded to walk out of the stable was if the depression could be lifted and he could be given back his will to live.

As luck would have it, Helen had just read a magazine article about registered aromatherapist, Caroline Ingraham, who works in conjunction with a homoeopathic vet. Helen contacted her with full details of the horse's condition. Helen received every help and support from her own vet, who is extremely open-minded and encouraged her to try anything she felt might help Cracker.

The aromatherapist asked for a hair sample, and after testing decided that she wanted to tackle the problem in three ways.

Firstly, using seaweed oil to detoxify the horse's system and support the liver and kidneys.

Secondly, using great mugwort oil, for its antihistamine action to deal with the allergic reaction.

Thirdly, using neroli oil for Cracker's emotional state, to help him deal with the sadness and depression he felt.

Before using the oils Helen was told to test the horse's enthusiasm for them by placing each one in turn near the horse's nostrils and watching his reaction. With all three oils he began to salivate and seemed to have a strong desire to get at them. Helen was a little suspicious that the horse was just being his usual greedy self, so tested a number of the other horses in the yard in the same way. All of these without exception either showed no interest at all or tried to evade the smell by turning their heads away.

Cracker was given 2 drops of the seaweed oil on a carrot twice a day for 2 days, and then once a day until he started to show a dislike for it (indicating that his body no longer needed it). This occurred within about 10 days.†

In addition to the seaweed oil, Helen used the solution made up of 30ml of walnut oil that Caroline had mixed with 5 drops of neroli and 5 drops of mugwort. This was gently smoothed on the area around his nostrils once a day.

Within one week the horse's mental and physical condition had improved beyond recognition – he was buoyant and eager to leave his stable, and a week later was pronounced fully fit by the vet. I am pleased to say that twelve months further on the horse has remained fit and healthy and is getting up to all his old tricks.

---

† *NB: Professional advice should always be sought when using essential oils especially when they are being administered orally.*

Homoeopathy

By Tim Couzens BVetMed, MRCVS, VetMFHom

Homoeopathy dates back to Greek times and its name derives from two words: *homoios* meaning 'like' and *pathos* meaning 'suffering'. The full potential of homoeopathy was not realised until the late eighteenth century and the pioneering work carried out by the German physician Samuel Hahnemann.

Disillusioned with the medical approach used by his colleagues, Hahnemann undertook translation work. Whilst working on Cullen's *Materia Medica* (a medical book listing treatments of the day), Hahnemann found himself in disagreement with the proposed action of one particular remedy, cinchona bark – this was one of the few effective medicines of the time and used in treating malaria. Over the course of several days he took repeated doses of cinchona and surprisingly developed a set of symptoms closely resembling those of malaria. Once he stopped taking the medicine the symptoms disappeared and he returned rapidly to good health. His conclusion from this simple experiment was that 'like could cure like'. Fundamentally if a substance could cause certain symptoms, it could be used to cure a patient displaying similar signs. This is the basis upon which the whole of homoeopathy is founded.

Over the next twenty years Hahnemann continued to test this theory using both himself and his friends as 'guinea pigs'. He carefully noted down the symptoms, both physical and mental, which developed as a result of taking each individual substance he investigated. These he compiled into a Materia Medica. By listing what a substance could cause, he was aware of what it was capable of curing in terms of symptoms. In the winter of 1812/13 he was able to put his theories into practice by successfully treating a typhoid outbreak.

Hahnemann was aware that some of the remedies he tested were toxic, so he set about finding the minimum dose needed for a cure. He began diluting his remedies, finding surprisingly that as he did so, the more effective they became. This is of course in direct contrast to modern drugs, which become less effective when diluted. His discovery, that diluted substances could be used to heal, meant that some of the most poisonous substances could be put to use therapeutically, without worrying about and side-effects or toxicity.

This meant that almost any substance can be prepared homoeopathically – even metals, minerals, snake venoms, and diseased tissues – and then used to heal in their homoeopathic form. Preparation of remedies involves serial dilutions with vigorous shaking in between each individual stage, a process known as succussion, which is vital to the correct preparation of the remedy.

As an example Arnica 30c, a remedy frequently used for injury and shock, would be prepared initially from an alcoholic extract of the arnica plant, known as the 'mother tincture'. One drop of this would be added to ninety-nine drops of an alcohol/water mixture and the test tube shaken vigorously, or succussed. The contents of this tube contain the first centesimal potency, or 1c dilution. One drop of this would then be added to ninety-nine drops of alcohol/water mixture and once again shaken to give the 2c potency. This process would be repeated a further twenty-eight times until the 30c potency was achieved. A few drops of the required potency can then be added to lactose tablets or powders, which then become potentised with the active remedy.

If we examine the process scientifically, the dilution at the 6c potency is equivalent to a dilution of 1:1,000,000,000,000, more or less equivalent to one drop of mother tincture in the combined volume of fifty swimming pools. One of the main problems is that at these dilutions it is difficult to see how the remedies can act, especially since past the 12c potency there will not be any of the molecules of the original substance left. These points form the main basis on which the critics of homoeopathy claim that it is unscientific, does not work and any result is due largely to a natural cure or a placebo effect. Yet homoeopathy remains effective in treating a wide variety of problems in animals, where the placebo effect can be discounted and the number of successes lies outside the realms of chance.

No one knows for sure how homoeopathic remedies work. Much research is currently underway. One popular theory is that each individual potentised remedy contains a form of energy, derived from the original substance through the process of dilution and succussion and stored as an 'imprint' on water molecules. It is this energy which heals, acting possibly through the immune system, correcting imbalances in the body which manifest themselves as illness.

Correctly applied, homoeopathy can be a remarkable and effective system of medicine which is both safe and without side-effects. The key to successful use of homoeopathy lies in matching the animal's symptoms to the remedy fairly accurately. The closer the match, the more likely the chance of a response. If, by chance, the wrong remedy is given then nothing happens, but neither is any harm done.

Remedies are available in various forms: tablets (which come in soft or hard types), granules, powders, and drops. They need special care when being handled and stored. They are easily de-activated by strong smells and vapours such as camphor and eucalyptus. Consequently, they are best stored away from any aromatherapy products. Bright light, heat and electromagnetic radiation (e.g. from televisions, mobile phones and computers) can de-activate or depotentise them, which will render them useless. Stored properly the remedies will keep for many years without losing their efficacy.

Most people with horses use tablets because they are the most readily available and the cheapest. The one drawback with tablets is the fact that they should not be handled as this can depotentise them. To administer tablets: carefully tip the remedy into the bottle lid and from there into the horse's mouth. If this is not possible (we all know how difficult some horses can be) then tip the tablet into a hole made in a small piece of carrot or apple and feed to the horse.

Choice of remedy and dosage is important. This information along with detailed advice can be obtained by contacting a homoeopathic vet.

For a list of qualified homoeopathic veterinary surgeons in Britain, write to the British Association of Homoeopathic Veterinary Surgeons, whose address can be found on page 166.

Radionics

When people talk about the 'black box' or 'being on the box' they are referring to the science of radionics. This is an alternative therapy which many people have difficulty understanding and 'getting to grips with'. The basic principle of radionics was first discovered by an American physician, Dr Albert Abrams, who was born in 1863. His theories were later investigated in the early 1900s and additional research was carried out in England in the 1940s. The whole concept of this therapy is to restore the body to a harmonious and balanced state in tune with nature, and to encourage the animal (or person) to 'heal itself'.

The therapy is based on energy patterns and the disharmony that occurs in these patterns when an animal or human is ill. The symptoms of the illness only tell a small part of the story and the radionics practitioner's aim is to get to the cause of the problem and to view the body as a whole, both mentally and physically.

The practitioner, by directing corrective energy patterns through the radionic instruments, brings the 'patient's' patterns back into balance, allowing the animal or person to 'heal itself'.

One of the advantages of radionic treatment is that it can be carried out at a distance, and does not require the presence of the horse. This is particularly advantageous to horse owners, especially in cases of emergency, when immediate action can be taken by the owner just telephoning the practitioner.

When you first approach the practitioner he/she will ask for a completed case history, which should include full details of any problems along with details of any medication or treatment the horse is currently receiving. This allows the practitioner to build up a complete mental picture of the horse. In addition to this the practitioner will ask for a 'witness'. This is usually a piece of mane or tail hair, which will act as a link between the radionic instruments and the horse.

The practitioner, using the hair sample, then makes an analysis of the horse by scanning the functioning of its systems on a radionic instrument, measuring and recording any malfunctions, infections, injuries or blockages. Once the practitioner has identified the systems which are at fault, he will try to find the cause, be it physical or psychological.

In conjunction with the radionic therapy, the practitioner may suggest the use of other treatments, including homoeopathy, herbs, physiotherapy, osteopathy, aromatherapy, etc.

It is important that the horse owner keeps the practitioner up to date with the horse's progress and any changes in the horse's condition. This will allow the treatment to be altered accordingly.

Animals seem to respond well to radionics, which can be successful in the treatment of allergies, infections, wounds and fractures, as well as mental problems such as fear and jealousy. Many people who use radionics practitioners for their horses may even keep them 'on the box' for a few months after they have sold them to a new owner, by way of keeping the horse under observation during this sometimes stressful time.

For further information on the history of radionics, details of training courses, and a list of qualified practitioners contact the Radionic Association, whose address can be found on page 167.

Chiropractic

The name chiropractic means 'done by the hand' and is derived from the Greek words *cheir* meaning 'hand' and *praktikos* meaning 'done by'.

In simple terms, chiropractic treatment is a way of treating disorders of the body by viewing the body as a whole, examining the skeletal structure and its associated musculature, determining where the problem lies, and then performing the required manipulation.

The chiropractic treatment of horses, dogs and farm animals has become much more widely used and recognised as an effective alternative treatment in recent years. Many top flat racing, national hunt, show jumping, dressage and eventing trainers now use chiropractors as a matter of course, both to prevent and correct problems.

Horses in particular can have spectacular falls when being ridden, and even whilst out in the fields. One mare I knew of had been cantering around her field and slipped over, crashing into a post and rail fence. The mare got to her feet and, apart from some superficial cuts, appeared to be none the worse for wear. As fate would have it she had been covered and was in foal; she was not ridden after the fall and went on to produce a healthy foal. It was not until the foal had been weaned and the owners started to bring the mare gently back into work that they noticed a dramatic change in her personality. Previously sweet natured and easy to handle, the mare became aggressive, difficult to handle and eventually dangerous. The vet was called and the horse examined, but no reason could be suggested for the sudden change of temperament.

During a conversation with the owners I asked if the mare had been involved in an accident and they remembered the fall of eighteen months previous. It was possible that the mare had damaged herself in the fall and that the problem had been compounded by carrying the foal. This damage and the resulting pain and discomfort it caused had not become apparent until she was ridden again a year and half after the original injury.

With the consent of the mare's vet, a chiropractor was called in and a full examination of the horse's musculo-skeletal system was carried out. It was found that indeed there had been damage done to the sacro-iliac area, causing mis-alignment. This damage had been made worse by both carrying and bearing the foal.

Several sessions of manipulation and massage were given to the horse along with a period of time off, followed by gentle work. Almost immediately the mare's behaviour changed; she regained her usual sweet nature, lost her aggressive attitude and was no longer difficult to handle. Within a few months she was back in full work, and has since had no repetition of the problem.

Very often a dramatic change of temperament along with uncharacteristic behaviour like bucking, rearing, napping, and even head-shaking can be an indication of pain in the back. More veterinary surgeons are now accepting the benefit of working closely with people such as chiropractors, osteopaths, sports therapists, etc. and will happily refer horse owners to these practitioners after they have first examined the horse to ensure the condition is not serious or life-threatening.

McTimoney Chiropractic

John McTimoney is believed to be the first chiropractor to formulate a chiropractic analysis and treatment specifically for animals. McTimoney chiropractors work in conjunction with veterinary surgeons and there are now many more vets who are prepared to refer appropriate cases to McTimoney animal chiropractors.

For further information on McTimoney Chiropractic and a directory of registered practitioners, contact the McTimoney Chiropractic Association, whose address is given at the back of the book.

Other Therapies and Treatments

In addition to the alternative therapies and treatments already covered in this section there are a number of other 'alternative healing' options open to horse owners. Unfortunately I do not have the knowledge or experience to explain these various methods in detail. Suffice to say that I have had personal experience of the benefits to be gained by using them and would urge horse owners to keep an open mind when considering their use. Many horses that have been 'written off' and would by now be dead, owe their lives to the determination of their owners to explore all avenues of healing before making the 'final decision'.

These include: acupuncture, reflexology, cranial osteopathy, applied kinesiology and healing.

Glossary

A

Abortifacient – will induce an abortion.

Alterative – this is an old-fashioned term, which has more recently been superseded by the term 'blood purifier'. Alteratives will speed up tissue renewal.

Analgesic – pain relieving.

Anaphrodisiac – reducing sexual desire.

Anodyne – something that reduces pain and irritation.

Antacid – a substance that reduces acidity, particularly in the gut.

Anthelmintic – an agent that will cause the death, elimination or expulsion of worms and parasites.

Antibacterial – helps the body resist or destroy pathogenic micro-organisms.

Antibiotic/anti-infective – having the action of helping the body withstand infection or infestation. Note: some essential oils and cider vinegar have this action.

Antifungal – helps the body kill or inactivate fungi or fungal infections.

Antihistamine – having a neutralising effect on the body's release of histamine.

Antihypothyroid – having an action on thyroid deficiency.

Anti-infective – *see* Antibiotic.

Anti-inflammatory – any remedy that reduces inflammation in the body, although herbs tend to work more by aiding the inflammation to cleanse the area, rather than suppressing the inflammatory process.

Antimicrobial – helps the body resist or destroy micro-organisms such as fungi and bacteria.

Antiphlogistic – relieving pain and inflammation.

Antirheumatic – having the action of preventing, relieving or curing rheumatism.

Antiseptic – having the action of destroying or inhibiting the growth of bacteria and other micro-organisms.

Antispasmodic – reducing spasm or tension particularly in areas of

Glossary

smooth muscle, such as gut walls, bronchial tubes, etc.

Antitussive – suppresses or relieves coughing.

Antiviral – something that is effective against viral infection.

Aperient – a mild laxative promoting natural bowel function, e.g. linseed, rosehips.

Aphrodisiac – stimulates sexual desire and functions.

Aromatic – having a distinctive fragrant smell.

Astringent – having a binding or contracting action on mucous membranes and tissue, usually because of the presence of tannins.

C

Calmative – calming, sedative.

Carminative – eases flatulence and colic in the gut.

Cholagogue – will stimulate the production and flow of bile, which as a result will have a slightly laxative effect on the digestive system.

D

Demulcent – having the action of soothing and protecting membranes, notably in the gut, mouth, throat, urinary system, skin and wounds. Normally due to the mucilage content which has a characteristic 'slimy' consistency.

Diaphoretic – will promote and increase sweating to help reduce body temperature in fevers.

Diuretic – to provoke an increase in the flow of urine

E

Electrolytes – the natural salts, e.g. sodium chloride, sodium bicarbonate, potassium chloride, that are present in solution in the body. Loss of electrolytes through sweating, for example, can cause dehydration and exhaustion.

Emmenogogue – *see* Abortifacient.

Emollient – having the action of soothing the surface to which it is applied.

Expectorant – having the action of encouraging the passage of mucus up the bronchials.

F

Febrifuge – reduces or prevents fever.

Flavonoids – plant constituents with a variety of properties, including diuretic, antispasmodic, anti-inflammatory and antiseptic. Certain flavonoids, such as rutin, have the effect of healing and strengthening peripheral blood vessels.

G

Galactogogue – will encourage the production of milk.

H

Haematoma – an accumulation of blood within the tissues that clots to form a solid swelling.

Haemostatic – will stop or control bleeding.

Hepatic – will support, stimulate and in some cases protect the liver.

Glossary

Hepatoprotective – having the action of supporting and protecting the liver.
Holistic – viewing the body as a whole, and having consideration for its physical, mental, environmental, cultural and spiritual aspects.
Hypotensive – having the action of reducing high blood pressure.

I

Immuno-stimulant – supports and stimulates the body's immune systems.

L

Laxative – to promote bowel movements.

M

Mucilage – will act locally as a demulcent to soothe and protect mucous membranes.

N

Nervine – restoring the nerves.

O

Oedema – excessive accumulation of fluid in the body tissues, usually caused by an injury or inflammation.
Oestrogenic – stimulates female hormones.

P

PMS/PMT – pre-menstrual syndrome or pre-menstrual tension.
Parasiticide – a substance that kills parasites.
Parturient – herbs that will help in birthing.
Pathogens – micro-organisms, such as bacteria, which live as parasites in the body of an animal, plant or man and cause a disease.
Phytotherapy – a term more widely used in Europe, but now being adopted by the UK and the USA to describe the use of plant remedies in medicines.
Prophylactic – having the action of protecting or preventing the body from contracting illness and disease.
Pulmonary – relating to or affecting the lungs.

R

Rubefacient – an external stimulant that has the action of creating gentle local irritation to the skin, stimulating the blood vessels to dilate, thereby increasing the local circulation and enhancing the dispersal of toxins and waste products.

S

Sedative – reducing nervous tention or excitement; sedating.
Spasmolytic – having the action of reducing spasms of smooth muscles, such as the intestines and bronchials.
Stimulant – energy producing; will enliven and stimulate the bodily functions.
Stomachic – good for the stomach.

V

Vasodilator – an agent that causes

Glossary

widening of the blood vessels.
Vermifuge – worm expellant action on the intestine.
Volatile oil – contained in many plants, the oil can be vapourised with heat.
Vulnerary – used to heal wounds.

REFERENCES

1. *Gastric Ulceration in Mature Thoroughbred Horses* C.J. Hammond et al. (1986) *Equine Veterinary Journal* 18(4), 284-287.
Investigation of the Number and Location of Gastric Ulcerations in Horses in Race Training (Submitted to the California Racehorse Postmortem Program) Johnson et al. (1994) *American Association of Equine Practitioners 40th Annual Convention Procedings.*

2. *Practical Veterinary Pharmacology, Materia Medica and Therapeutics* Howard Jay Milkes DVM (Bailliere, Tindall and Cox, London, 1937)
A Manual of Veterinary Therapeutics and Pharmacology E. Wallis Hoare FRCVS (Bailliere, Tindall and Cox, London, 1895)
Veterinary Medicines, Their Actions and Uses (12th Ed.) Finlay Dun (David Douglas, Edinburgh, 1910)
Special Pathology and Therapeutics of the Diseases of Domestic Animals Hutyra and Marek (Bailliere, Tindall and Cox, London, 1926)
Veterinary Posology (2nd Ed.) G.A. Banham FRCVS (1901)
A Concise History of Veterinary Medicine D. Karasszon (Akedemia Kiado, Budapest, 1988)
The Horse W. Youatt (Longman, Green, Longman and Roberts, 1859)
Veterinary Materia Medica and Therapeutics (5th Ed.) Winslow (Bailliere, Tindall and Cox, London, 1907)
• All the above mentioned books can be found in the Royal College of Veterinary Surgeons' library in London.

3. Wagner, H. in *Advances in Chinese Medicinal Materials Research* Eds. H.M. Chang et al, (World Scientific, Singapore, 1986)

4. Brandt, L., (1967) Scand. J., *Haematol.* Suppl. 2

5. *Horses and Homeopathy, A Guide for Yard and Stable* Mark Elliott BVSc, MRCVS and Tony Pinkus BPharm, MRPharmS. (Ainsworths

References

Homoeopathic Pharmacy, London, 1994)

6. *The Magic of Green Buckwheat* Kate Spencer (Romany Herb Products Ltd, York, 1987)

7. *Pharmacology and Applications of Chinese Materia Medica, Vol 1* Ed. H. Chan and P. But (World Scientific, Singapore, 1986)

8. Furuya, T. and Araki, K. (1968) Studies on Constituents of Crude Drugs 1 *Chem. Pharm. Bull.* 16 2512-2516

9. *The Case for Comfrey* article by Penelope Ody in 'Herbs', Spring 1993

10. *Encyclopedia of Common Natural Ingredients used in Food Drugs and Cosmetics* Albert Y. Leung (John Wiley and Sons Inc., New York, 1980)

11. Lanhers M.C. et al, (1992) 'Anti-Inflammatory and Analgesic Effects of an Aqueous Extract of *Harpagophytum procumbens*' Planta Medica

12. *Echinacea – The Immune Herb* Christopher Hobbs (Botanica Press, Capitola, California, 1990)

13. *A Modern Herbal* Mrs M. Grieve (Penguin Books, London, 1980)

14. *Eleutherococcus Preparations in Animal Breeding* T.A. Lyaputsina (Kolos Publishers, Moscow, 1980)

15. *Folk Medicine* D.C. Jarvis MD (Pan, London, 1960)

BIBLIOGRAPHY & FURTHER READING

Listed below are a selection of books that I have found helpful over the years. Some of them are herbal reference books, others contain invaluable information on the general care of horses and ponies. I have added relevant comments on some of the books.

British Herbal Medicine Association
British Herbal Pharmacopoeia
Volumes: 1, 2, 3, 4. (BHMA, Bournemouth, Dorset, 4th Ed. 1996)

British Herbal Compendium
Volume 1 – Companion to the British Herbal Pharmacopoeia Volume 1. (BHMA, Bournemouth, Dorset, 1992)

These books contains a series of monographs on a number of well known herbs. Although aimed at the application of herbs in human herbal medicine, they offer a great deal of information to the interested lay person along with details on recent research.

Pasture Management for Horses and Ponies
Gillian McCarthy
(BSP Professional Books, Oxford, 1987)
A must for any horse owner, whether they own land or not. The book covers everything that a horse owner needs to know, from the care and management of grassland, through fencing suggestions, soil types, drainage, fertilising, hay crops, field shelters, parasite control and many more topics.

BIBLIOGRAPHY

Horses and Homoeopathy, A Guide for Yard and Stable (1994)
Mark Elliot BVSc, MRCVS and Tony Pinkus BPharm, MRPharmS
Available through:
Ainsworths Homoeopathic Pharmacy,
38 New Cavendish Street, London, W1M 7LH
Tel: 0171 935 5330

US Distributor:
Echo Publishing Inc., Rt. 9 Box 72-8, Santa Fe, New Mexico 87505
Tel: (505) 989 7280

A useful and easy-to-understand guide to the use of homoeopathy for horses. Explains the majority of common problems with horses and suggests which homoeopathic remedy to use.

Garlic for Health
David S. Roser MCIM
(Martin Books, Simon and Schuster Consumer Group, Grafton House, 64 Maids Causeway, Cambridge, CB5 8DD, 1995)
A fascinating book written by the man who probably knows more about garlic and its health applications than anybody else. David Roser is the head of the Garlic Research Bureau, which publicises the benefits of garlic for health. The book includes details on the use of garlic for animals.

Folk Medicine
D.C. Jarvis, MD
(Pan Books, London, 1960)
A fascinating insight into the folk medicine practised by the farming people of Vermont USA, who, because of their isolation, were forced to find their own natural remedies for many common ailments.

Aromatherapy an A-Z
Patricia Davis
(C.W. Daniel Co. Ltd., Saffron Walden, 1988)
One of the aromatherapy 'bibles'; easy to navigate.

Bibliography

The Art of Aromatherapy
Robert Tisserand
(C.W. Daniel Co. Ltd., Saffron Walden, 1977)
A general study of the subject with details of a wide selection of essential oils and their application.

The Complete Herbal Handbook for Farm and Stable
Juliette de Bairacli Levy
(Faber and Faber, London, 1988)
Regarded as the 'bible' for animal owners. Certainly the classic of its time.

From Foal to Full-Grown
Janet Lorch
(David and Charles, Newton Abbot, 1995)
A book I have found invaluable as an easy-to-understand and comprehensive guide to breeding and foaling.

A Modern Herbal
Mrs M. Grieve
(Penguin Books, London 1980)
First published in 1931, *A Modern Herbal* is the complete and comprehensive herbal of its time, and covers all but the more recently discovered species or uses. Definitely a herbal 'bible', and it makes reference to use with animals.

Echinacea – The Immune Herb
Milk Thistle – The Liver Herb
Christopher Hobbs
(Botanica Press, Capitola, California, 1990, 1992)
Two excellent books packed with information and research details. Excellent additions to anyone's herbal library.

The New Holistic Herbal
David Hoffmann
(Element Books, Shaftesbury, Dorset, 1990)
An excellent herbal for humans that gives detailed information on the preparation of herbs, including tinctures and extracts.

Bibliography

Woman Medicine – Vitex Agnus-Castus (1992)
Simon Mills MA, FNIMH
Amberwood Publishing
Braboeuf House, 64 Portsmouth Road, Guildford, Surrey
Tel: 01483 570821
Fax: 01483 282321
Simon Mills is vice-chairman of the British Herbal Medicine Association and one of the leading medical herbalists in the UK. He is a joint director of the Centre for Complementary Health Studies at Exeter University, and is responsible for much of the research carried out there. He has written several excellent books, including this one which is a must for any woman.

Potter's New Cyclopaedia of Botanical Drugs and Preparations
R.C Wren FLS
(C.W. Daniel Co. Ltd., Saffron Walden, 1989)
An excellent reference book for both the scientist and the layman.

The Complete New Herbal
Richard Mabey
(Penguin Books, London 1988)
A lovely book packed with information on herbs and their uses, both medicinal and culinary. It contains beautifully photographed colour plates of literally hundreds of plants. This book is particularly useful for those wanting to accurately identify plant species.

USEFUL ADDRESSES

Recommended Suppliers

UNITED KINGDOM

Hilton Herbs Ltd
Downclose Farm
North Perrott
Crewkerne
Somerset
TA18 7SH
Tel: 01460 78300
Fax: 01460 78302
Suppliers of individual herbs, tinctures, homoeopathic products, organic cider vinegar, and a full range of complementary health products for horses.

Hambledon Herbs
Court Farm
Milverton
Somerset
TA4 1NF
Tel: 01823 401205
Suppliers of top quality organically grown herbs.

Fiddes Payne Herbs and Spices Ltd
Pepper Alley
Unit 3B
Thorpe Way
Banbury
Oxfordshire
OX16 8XL
Tel: 01295 253888
Fax: 01295 269166
A useful source for smaller quantities of top quality herbs and spices.

Paul Richards Herbal Supplies
The Field
Eardisley
Herefordshire
HR3 6NA
Tel: 01544 327360
Produces organically grown comfrey products, including ointments and oils. Also hypericum and calendula oils.

USEFUL ADDRESSES

The Fragrant Earth Company Ltd
PO Box 182, Taunton
Somerset
TA1 1YR
Tel: 01823 335734
Fax: 01823 322566
Suppliers of top quality essential oils. Also stockists of nebulising diffusers for stable use.

Probiotics International
Matts Lane
Stoke-Sub-Harndon
Somerset
TA14 6QE
Tel: 01935 822921
Fax: 01935 826300
This company holds a Royal warrant, and produces a range of probiotics for most species.

Cotswold Grass Seeds
The Barn Business Centre
Great Rissington
Cheltenham
Gloucestershire
GL54 2LH
Tel: 01451 822055
Suppliers of herb seeds for sowing into paddock grass leys.

Henry Doubleday Research Association
The National Centre for Organic Gardening
Ryton Gardens
Ryton-on-Dunsmore
Coventry
CV8 3LG
Tel: 01203 303517

Cornish Calcified Seaweed
Newham
Truro
Cornwall
TR1 2SU
Tel: 01372 78878
Suppliers of seaweed for paddock dressing.

Equine Marketing
Ledgemoor
Weobley
Hereford
HR4 8QH
Tel: 01544 318196
Fax: 01544 318770
Manufacturers and distributors of top quality probiotics. Equine Marketing will also carry out faecal sample analysis.

Ainsworths Homoeopathic Pharmacy
38 New Cavendish St
London
W1M 7LH
Tel: 0171 935 5330
Fax: 0171 486 4313
An excellent source of all homoeopathic remedies, tinctures, ointments and homoeopathic supplies.

Forbes Copper
Garston House
Sixpenny Handley,
Salisbury
Wiltshire
SP5 5PB
Tel/Fax 01725 552300
Copper products for horses including pastern bands.

USA

**Echo Publishing Inc
(The Chamisa Ridge Catalog)**
Rt. 9 Box 72-8
Santa Fe NM 87505
Tel: (505) 989 7280
This company produce an excellent catalogue of 'natural and alternative' products for horse and rider. They also distribute Hilton Herbs products throughout the USA.

San Fransisco Herb Co
250 14th Street
San Francisco CA 94103
Tel: (800) 227 4630
This company sells a wide range of individual herbs for horses by mail order.

Alternative & Complementary Therapies

UNITED KINGDOM

The British Herbal Medicine Association
PO Box 304
Bournemouth
Dorset
BN7 6JZ
This association was founded in 1964 to advance the science and practice of herbal medicine in the UK.

The Herb Society
134 Buckingham Palace Road
London
SW1W 9SA
Contact the Society and enclose an s.a.e. for a list of veterinary surgeons who use herbal medicine.

British Association of Homoeopathic Veterinary Surgeons
Chinham House
Stanford-in-the Vale
Faringdon
Oxfordshire
SN7 8NQ.
Tel: 01367 710324
Contact the Secretary and enclose an s.a.e. for a list of homoeopathic veterinary surgeons.

The McTimoney Chiropractic Association
The Administrator
21 High Street
Eynsham
Oxford
OX8 1HE
Tel: 01865 880974/5

Useful Addresses

The Radionic Association
The Secretary
Baerlein House
Goose Green
Deddington
Banbury
Oxfordshire
OX15 0SZ
Tel/Fax: 01869 338852

Mrs A.L.G.Dower
Radionics Practitioner
Swinbrook Cottage
Swinbrook
Burford
Oxford
OX18 4DY
Contact: Mrs A.L.G
Dower FKCOLLR MRR

The Register of Qualified Aromatherapists
P.O. Box 6941
London
N8 9HF
Tel: 0181 341 2958

Caroline Ingraham MRQA
Essential Oils for Equines
c/o Fragrant Studies
Unit 1, Belvedere Trading Estate
Taunton
Somerset
TA1 1BH
Tel: 01243 378035
A fully qualified and registered aromatherapist who works in conjunction with veterinary surgeons.

Bach Flower Remedies Customer Enquiries
Broadheath House
83 Parkside
Wimbledon
London
SW19 5LP

The Dr Edward Bach Centre
Mount Vernon
Sotwell
Wallingford
Oxon
OX10 0PX
Tel: 01491 834678
This centre is where Dr Edward Bach lived, worked and developed his healing remedies. It is now run by a group of custodians who are dedicated by legacy to maintain the methods and use of the remedies as intended by Dr Bach. It was from fields, hedgerows and woods surrounding this property that Dr Bach collected his flowers to prepare the healing remedies. These same locations are still used as the source of the flowers for the Bach flower remedies.

USEFUL ADDRESSES

USA

American Herb Association
Box 353
Rescue CA 95672

California School of Herbal Studies
PO Box 39
Forestville CA 95436
Tel: 707 887 7457

Herb Research Foundation
1007 Pearl St. Suite 200
Boulder CO 80302

American Herbalists Guild
PO Box 1683
Sequel CA 95073

American Botanical Society
PO Box 201660
Austin TX 78720

International Veterinary Acupuncture Society
268 West Third
PO Box 2074
Boulder CO 80456

AUSTRALIA

National Herbalists Association of Australia
27 Leith Street
Coorparoo
Queensland 4151

Index

Page references in **bold** type refer to main entries.

Aaron's Rod *see* golden rod
Abrams, Albert 150
Absinthe *see* wormwood
Achillea millefolium see yarrow
aconite 129
acupuncture 95, 116, 153
adrenalin 66
afterbirth expulsion 51, 58, 91
agnus castus **25-6**, 99-100, 124-5
Agropyron repens see couch grass
alfalfa 102, 108, 114, 119
allantoin 32-3, 139
allergies
  respiratory *see* hay fever/ seasonal allergies; mucus
  skin *see* dermatitis; eczema; greasy heels; mud fever; rain scald; sweet itch
allicin 70
Allium sativum see garlic
alterative herbs *see* blood purifiers
Althea officinalis see marshmallow
American Purple Coneflower *see* echinacea
amino acids 66, 97
Amoracia rusticana see horseradish
anaemia 49, 50, **73**, 80
analgesic action 31, 36, 44, 102, 133
anaphrodisiac 25, 124
anger 96
angleberries *see* sarcoids
Anise *see* aniseed
aniseed **26**, 85, 95, 108, 112,

aniseed cont.
  118, 138
anthelmintic action 78, 92
anti-allergic action 95
anti-asthmatic action 95
antibacterial action 49, 54, 111, 112
antibiotic action 35, 40, 55, 67, 69, 70, 87
antibiotic alternatives **69-70**
antibiotics, conventional 68, 69-70, 81, 112
antifungal action 30, 54, 67, 125
antihistamine action 95, 132
anti-inflammatory action 31, 36, 45, 47, 61, 83, 102, 103, 115, 123, 133
antimicrobial action 69, 70, 77, 87, 92, 106, 107, 121, 126, 132
antiparasitic action 49
antirheumatic action 122
antiseptic action 44, 47, 58, 64, 70
antispasmodic action 79, 88, 117-18
anti-tumour action 29, 52, 80, 126
antiviral action 67, 112
anxiety 59, 75, 93, 114
aphrodisiacs 39, 91
apiol 30
Apium graveolens see celery seed
appetite
  insatiable 109
  see also mineral/vitamin deficiency; worms
  poor appetite **74**

appetite stimulants 36, 39, 42, 43, 62, 63, 85, 88
Apple Mint *see* mint
Arctium lappa see burdock
Arctostaphylos uva-ursi see uva-ursi
argent nit. 88
arnica **27**, 76, 102, 116, 123, 124, 129, 133, 139-40, 149
Arnica montana see arnica
aromatherapy 137, **144-7**
arsenicum album 132
Artemisia absinthium see wormwood
arthritis *see* DJD
aspirin 47, 61
asthma *see* hay fever/seasonal allergies
astringent action 47, 53, 57, 64, 77, 128
azoturia 65, 84, 114

Bach, Edward 143
Bach Flower Remedies 75, 88, 90, 96, 118, **143-4**
  Rescue Remedy 75, 88-9, 91, 118, 129, 132, 139, 143
  Rock Rose 90, 118, 143-4
bacterial infections 37, 87
Bardane *see* burdock
barley water 81
Barosma betulina see buchu
Bearberry *see* uva-ursi
Beargrape *see* uva-ursi
Beggar's Button *see* burdock
Bellis perennis see common garden daisy
bile production 35, 48, 49, 85, 105
biotin 53, 97

169

INDEX

Bird's Foot *see* fenugreek
Black (Mitcham) Mint *see* mint
Black Sampson *see* echinacea
Bladderwrack *see* kelp
bleeding *see* epistaxis
blood pressure, normalising 30, 42, 104
blood purifiers 30, 35, 40, 49, 54, 64, 77, 80, 86, 91, 100, 102, 115, 119, 121, 123, 131
blood vessels, burst *see* epistaxis
body salts 84
boils 29, 56
bone damage 102
bone pain 28
boneset (*Eupatorium perfoliatum*) **28**, 95, 101
box rest/stall rest **74-5**, 132, 133
bran 20
Briar Rose *see* rosehip
Bridewort *see* meadowsweet
bronchial complaints 95
bronchial spasms 59
Bruisewort *see* comfrey; common garden daisy
bruising 27, 30, 33, 61, **76**, 103, 107, 130, 139-40
buchu **28**, 81
buckwheat **28-9**, 76, 83, 87, 89, 95, 97, 102, 106-7, 108, 115, 117, 119, 123, 130, 132, 133
bugleweed 93
burdock 17, **29**, 77, 80, 86, 101, 102, 104, 105, 106, 110, 115, 119, 121, 123, 125, 126-7, 136, 138
burns 30, 44, 65

cabbage leaves 76, 102
calcification of joints and arteries 64
calcium 29, 35, 39, 97, 110
calendula 20, **29-30**, 31, 76, 77, 85, 86, 92, 93, 101, 102, 106, 107, 108, 110, 113, 114, 115, 121, 123, 125, 126, 127, 130, 131, 132, 133, 136, 137, 139, 140
Calendula officinalis see calendula
calming action 30, 50, 51, 52, 59, 65, 75, 96
camphor 54
capillary strengthening 28, 29, 54, 89, 95
castor oil 77, 125
catarrh 38, 41, 42, 49, 51, 58, 95, 111
Catch-weed *see* clivers
celery seed **30**, 83, 102, 115, 119, 123, 130
chamomile, German **31**, 75, 77, 79, 86, 88, 90, 96, 102-3, 107, 110, 115, 117, 123, 131, 135, 137, 140
Chaste Tree *see* agnus castus
Chasteberry *see* agnus castus
chiropractic **151-3**
cholesterol levels 39
cider vinegar **64-5**, 77, 83, 100, 107, 108, 110, 112, 117, 119, 122, 125, 130, 132, 133, 138, 140
circulatory stimulants 28, 41, 42, 43, 49, 54, 63, 83, 102, 104, 107, 115, 119, 123, 130, 133
Cleavers *see* clivers
clivers 30, **31-2**, 58, 77, 80, 83, 86, 93, 97, 100, 101, 102, 103-4, 105, 106, 107, 110, 113, 114, 115, 119, 121, 123, 125, 126, 130, 131-2, 133, 136
Clives *see* clivers

clover 80, 91, 108, 110, 114
coat and skin condition 31, 35, **76-7**
external skin conditions *see* skin irritation
cod liver oil 83
colic 26, 33, 49, 58, 59, 61, **78-9**, 117, 138
impactive colic 79
spasmodic colic 46, 78-9, 137
colitis 33, 46, 56
colocynthis 78
comfrey 17, 18, 20, 21, **32-5**, 73, 75, 76, 79, 83, 85, 86, 95, 97, 102, 103-4, 107, 110, 113, 115, 116, 118, 119, 123, 124, 128, 130, 131, 132, 133, 134, 135, 138, 139
Common Comfrey 34
common garden daisy 76
Common Thyme *see* thyme
competition nerves 87, 88
conjunctivitis 38, 89
constipation 28, 35, 59
copper 50, 109, 110
copper deficiency 109
copper pastern bands 124
Corn Poppy *see* poppy
Corn Rose *see* poppy
cortisol 66
couch grass 28, **35**, 58, 81, 101, 107, 110, 115, 123, 130, 138
cough 26, 40, 42, 46, 51, 52, 58, 65, 101, 111, 112
see also hay fever/seasonal allergies; mucus
covering, difficulties during 98
cramps 59, 61
cranial osteopathy 153
Crataegus oxyacantha see hawthorn

Index

Culicoides midge 132
Cushing's Disease **79-80**
cuts and abrasions 30, 40, 57
cystic ovaries 98
cystitis 28, 30, 31, 35, 37, 46, 58, 63, **81**

dandelion **35-6**, 77, 81, 84, 86, 97, 100, 101, 103-4, 105, 106, 107, 110, 115, 121, 123, 130, 136, 138
dappling 49
debility, general 53, 70
dehydration **84**, 92
demulcent action 45, 79, 81, 85, 87, 95, 112, 118
dermatitis **85**
  see also eczema; greasy heels; mud fever; rain scald; sweet itch
dermatophilosis see greasy heels
Dermatophilus congolensis 113
detoxifiers 101
devil's claw **36**, 83, 102, 104, 107, 115, 117, 123, 130, 133
diaphoretic herbs 20
diarrhoea see scouring
diet
  cleansing diet 83, 107
  healthy diet 77
  mineral/vitamin deficiency 93, 96-7, 104, **108-10**
dietary fibre 49
digestive carminatives 26, 55
digestive problems 40, 46, 54, 55, 56, 58, 70, 74, 76
  see also specific ailments e.g. colic
digestive tonics 29, 30, 41, 42, 49, 62, **85-6**
diuretic action 30, 31, 35, 36, 47, 50, 81, 115, 123, 130
DJD (degenerative joint disease/arthritis/

DJD cont.
  osteoarthritis) 28, 29, 30, 33, 36, 44, 49, **81-3**, 119
dock 105, 107
Dog Rose see rosehip
Dog's Grass (US) see couch grass
dust allergies 95

echinacea **37-8**, 41, 70, 77, 87, 89, 101, 102, 106, 107, 112, 113, 121, 125, 126, 132, 133, 136, 140
Echinacea angustifolia, E. purpurea see echinacea
eczema 29, 52, **86**
electrolytes 36, 84
Eleutherococcus senticosus see Siberian ginseng
emollient action 46, 95
epilepsy 55, 59
epistaxis 28, 51, 61, 63, **86-7**
equine nutritionist 86
essential oils 54, 57, 75, 83, 88, 90, 112, 115, 123, **144-7**
eucalyptus 112, 140
Eupatorium perfoliatum see boneset
Euphrasia officinalis see eyebright
European Arnica see arnica
European Willow see willow
eventing and show jumping 120-1
excitability **87-8**
  see also hyperactivity; nervous gut; nervousness
exhaustion **88-9**
expectorant action 33, 40, 42, 45, 46, 51, 58, 87, 95, 111
eye problems **89**
  conjunctivitis 38, 89
  keratitis 89
  sore and inflamed eyes 30,

eye problems: sore cont.
  31, 38, 60, 61
  stinging 38
  ulceration 38, 89
  weeping 38, 95
eyebright **38**, 89, 96

Fagopyrum esculentum see buckwheat
Fairy Clock see dandelion
fasting 76-7
fear **90**
feed
  additives 39, 40, 49
  allergic reactions to 86, 100, 130
  antibiotics in 69
feeding programmes 86
fennel 108, 138
fenugreek **39**, 85, 91, 108, 110, 115, 123, 126
fertility 45, **90-1**
fevers 27, 28, 43, 47, 55, 61, 63
Feverwort see boneset
Filipendula ulmaria see meadowsweet
flatulence 49, 59, 85, 118
flavonoids 42, 49, 53, 54, 76, 98
fluid retention 31
  see also diuretics
foal scour **92**
foaling **91**
Foenugreek see fenugreek
folic acid 73
founder (US) see laminitis
Fucus vesiculosus see kelp
fungal infections 57, 70, 87, 125

Galium aparine see clivers
Garden Lavender see lavender
Garden Sage see sage
Garden Thyme see thyme

171

INDEX

garlic 21, 39, **40**, 70, 77, 80, 86, 87, 89, 91, 92, 95, 101, 104, 106, 107, 110, 111-12, 113, 121, 125, 126, 127, 131, 132, 133, 134, 136, 138, 139, 140
gastric acid secretions 45
gastric bleeding 47
gastric disorders 30, 33, 39, 45, 46
gastric ulcers 10, 30, 33, 36, 45, 46, 47, 56, 73, **134-5**
Gattefossé, René-Maurice 144
gelatin 98
gentiane 75
ginger 21, 43
ginseng 91, 101
glandular problems **92-3**
see also hormonal imbalances
glandular system 80, 92
Glycyrrhiza glabra see liquorice
Gold-bloom *see* calendula
golden rod **41**, 110, 139
Goose-grass *see* clivers
greasy heels **113-14**
greater celandine 137
Greek Hay-seed *see* fenugreek
Green Ginger *see* wormwood
Grieve, M. 39, 40
grooming 77
gut
  acidity in the 66, 117, 135
  bacterial flora 40, 66, 68, 70
  nervous gut **117-18**

Hagthorn *see* hawthorn
Hahnemann, Samuel 148
hair growth 21, 125
Hamamelis virginiana see witch hazel

Harpogophytum procumbens see devil's claw
Haw *see* hawthorn
hawthorn **41-2**, 83, 97, 102, 104, 107, 110, 115, 117, 119, 123, 130, 133
hay fever/seasonal allergies **94-6**
see also head-shaking; sweet itch
head-shaking 94, **96-7**
heart rate 41
heels, cracked *see* mud fever
Helmet Flower *see* scullcap
hepatic action 35, 77, 105, 106, 115, 121, 123
Herb of Grace *see* vervain
herbs
  compresses/fomentations 20
  decoctions 18-19
  dosages 15-16
  dry or fesh-cut 17
  infusions, teas and brews 16, 17-18
  ointments 21
  picking 12
  poultices 20-1
  preparation 17-21
  sources of 164-6
  tinctures, extracts, concentrates 16, 19
Hoarhound *see* horehound
Hogberry *see* uva-ursi
holistic approach to medicine 83, 117
homeopathy 21, 95, 103, 104, 107, 127, 132, 133, 136, 139, **148-50**
honey 21, **65**, 68, 70, 81, 88, 117, 128, 135
Hoodwort *see* scullcap
hoof growth 35, 53
hoof problems **97-8**
horehound 17, **42-3**, 85

hormonal imbalances 25-6, 39, 80, 93, **98-100**, 103
see also glandular problems; rigs and false rigs
horseradish 21, **43**
hyperactivity 93, **100**
hypericum 20, 21, **43-4**, 83, 92, 113, 115, 123, 132, 139
Hypericum perforatum see hypericum
hyperthyroidism 93
hypothyroidism 44, 73, 93
hysteria 55

immune system 106, 120
immunity problems 73, **100-1**
see also post-viral syndrome
immuno-stimulants 27, 37, 38, 70, 89, 95, 101, 112, 126
Indian Elm *see* slippery elm
infection, resisting 70, 80, 87, 89, 106
inflammation 27, 44, 56, 61, 79, 82, **102-3**, 107, 115
influenza 28, 43, 55
inhalants 49, 57, 112
insect bites 40, 130, 131
insectidal lotions 62
intestinal inflammation and irritation 79
intestinal spasm 78
iodine 44, 93, 97, 110
iron 29, 35, 49, 50, 73, 80, 97, 110
Italian Liquorice *see* liquorice

jaundice 35
joint mobility 30, 82, 83, 102

Kansas Snakeroot *see* echinacea
kelp **44**, 73, 77, 80, 89, 93, 97, 102, 104, 106, 108,

INDEX

Kelp cont.
  109-10, 110, 115, 119, 121, 123, 131, 133, 134, 138
Kelpware *see* kelp
keratitis 89
kidney disorders 35, 84
kidney function 29, 77, 101, 115
kidney stones 41, 46, 50
kinesiology 153
Knight's Milfoil *see* yarrow
Knitback *see* comfrey
Knitbone *see* comfrey
Korean ginseng 91

Lady of the Meadow *see* meadowsweet
lameness 82, 102, 115, 133
laminitis 10, 31, 42, 49, 80, **103-4**
lavender **45**, 75, 83, 88, 90, 97, 102-3, 112, 115, 116, 118, 123, 124, 125, 133, 137, 140
Lavendula angustifolia see lavender
laxative action 60
legs, filled 31
Leopard's Bane *see* arnica
lethargy 73, 80, 88, 94, 104
lice 26, 40, 62
Licorice *see* liquorice
ligament strain or damage *see* tendon and ligament strain or damage
linoleic acid 48
linseed 79, 108
Lion's Teeth *see* dandelion
liquid paraffin 79
liquorice **45-6**, 79, 91, 101, 106, 110, 112, 118, 119, 135, 138
liver function 29, 36, 45, 48, 54, 77, 80, 101, 105, 115, 135

liver problems 35, 48, 60, 76, **105-6**, 138
Long Buchu *see* buchu
lucerne *see* alfalfa
Lycopus virginicus see bugleweed
lymphangitis 31, **106-7**, 130
lymphatic drainage 125
lymphatic herbs 123
lymphatic system 30, 31, 77, 80, 83, 93, 101, 106, 115, 126, 130

Mad-dog Skullcap *see* scullcap
Madweed *see* scullcap
magnesium 29, 35, 84, 96-7, 109, 110
malabsorption of nutrients 117, 118
Mallards *see* marshmallow
mane and tail growth 77
mane and tail rinses 49, 64, 77, 132
manganese 110
mange 62
mares
  fertility 45, **90-1**
  in foal 26, 36, 51-2, 54, 55, 62
  hormonal imbalances 25-6, 39, 80, 93, **98-100**, 103
  milk production 39, 48, 49, 50, 55, 60, **107-8**
  nursing mares 40, 50, 55
Marian's Thistle *see* milk thistle
Marigold *see* calendula
Marrubium vulgare see horehound
marshmallow 20, 28, **46**, 56, 58, 79, 81, 85, 87, 95, 97, 108, 112, 113, 118, 119, 135, 138
Marybud *see* calendula

Mary's Thistle *see* milk thistle
massage oils 33, 57, 145
mastitis 31, **107**
Matricaria recutita see chamomile, German
Mauls *see* marshmallow
May *see* hawthorn
McCarthy, Gillian 114
McTimoney chiropractic 152-3
meadowsweet **47**, 83, 97, 102, 106, 110, 113, 115, 117, 118, 123, 127, 128, 130, 131, 133, 135, 138
Melaleuca alternifolia see tea tree oil
mental confusion 73
mentha piperita see mint
menthol 49
midge bites 131, 132
Milfoil *see* yarrow
milk
  drying off 49, 55
  milk production 39, 48, 50, 60, **107-8**
milk thistle **47-8**, 80, 104, 106, 115, 123, 135, 138
mineral/vitamin deficiency 93, 96-7, 104, **108-10**
mint **48-9**, 85, 91, 118
Missouri Snakeroot *see* echinacea
Monk's Pepper *see* agnus castus
mood swings *see* hormonal imbalances
Moose Elm *see* slippery elm
Mountain Cranberry *see* uva-ursi
Mountain Daisy *see* arnica
mouth infections and inflammation 52, 55
mouth ulcers 52, 55, 60, **134**
mouth washes 134
mucilage 35, 46, 47, 56

173

INDEX

mucilagenous action 46, 79, 95
mucus 28, 38, 52, 57, **111-12**
mucus expulsion 40, 45, 46, 87
mud fever **113-14**
muscle damage, wastage or tension 27, 57, 59, 61, 98, 102, **114-16**
muscle rubs 33, 57, 59, 83
muzzle, spots on the 94
myrrh 103

nasal haemorrhage 86-7
nasal irritation 87, 96
navicular syndrome 28, 42, 63, **116-17**
nebulising diffusers 112
neroli 75, 147
nervine relaxants 79, 88, 90, 100, 117
nervine tonics 60, 88
nervous gut **117-18**
nervous stress, tension or exhaustion 55, 59, 75, 118
nervous system 76, 85, 88
nervousness 59, **118**
  see also excitability; fear; nervous gut
nettle 17, 30, **49-50**, 73, 76, 77, 80, 83, 86, 89, 93, 97, 101, 102, 103-4, 106, 107, 108, 110, 113, 115, 119, 121, 123, 130, 131, 134, 136
nettle rash 50
niacin 70
Nipbone *see* comfrey
Nose Bleed *see* yarrow

oak 75, 97, 128, 144
oats 88, 100, 101, 108, 110, 134
oedemas *see* soft swellings
oestrogenic activity 45, 52, 90, 91

old age **118-19**
Old Man's Pepper *see* yarrow
osteoarthritis *see* DJD
Oval Buchu *see* buchu

paddock management 109, 138
pain reduction 28, 107
panic attacks 90, 144
Papaver rhoeas see poppy
parasitic infections 26, 62, 73
  see also lice ; scabies; worms
parotid gland 93
parsley **50-1**, 73, 110
Pee-the-Bed *see* dandelion
pelvic muscles 90, 91
Peppermint *see* mint
peppermint oil 49, 85
Petroselinum crispum see parsley
phosphorus 110
photosensitivity 44
phytase 66
pigmentation loss 109
Pimpinella anisum see aniseed
Pin Heads *see* chamomile, German
pituitary gland 80, 98, 99
PMT/PMS *see* hormonal imbalances
poison, drawing 20, 56, 140
pollen allergies 94-5
pollution 10, 94, 100
poppy **51**, 88
post-viral syndrome 37, **120-1**
Pot Marigold *see* calendula
potassium 29, 35, 64, 84, 110
pregnancy 26, 36, 51-2, 54, 55, 62
Prickly Comfrey 34
probiotics **66**, 68, 70, 81, 88, 117, 119, 128, 135
prohibited substances 54, 146
propolis **67**, 113-14, 132

protein 49
pyrrolizidine alkaloids (PAs) 34

Quaker Bonnet *see* scullcap
Queen of the Meadow *see* meadowsweet
Quick Grass *see* couch grass
quidding 119

rabies 55
racing 120
radionics 150-1
rain rot *see* rain scald
rain scald **113-14**
raspberry **51-2**, 91, 110
red clover **52**, 77, 89, 100
Red Elm *see* slippery elm
Red Poppy *see* poppy
Red Sage *see* sage
reflexology 153
relaxant action 59, 86
respiratory problems 28, 33, 40, 41, 42, 45, 49, 58, 70, 73, 87, 94, 101, 111
restlessness 31, 59
rheumatism 27, 29, 33, 35, 44, 47, 49, 50, 54, 61, 63, 114, 119, **122-4**
  see also muscle damage, wastage or tension
riboflavin 53, 70, 110
rigs and false rigs **124-5**
ringworm 40, 57, 64, **125**
Rockberry *see* uva-ursi
Rockweed *see* kelp
Rosa canina see rosehip
rosehip **53**, 70, 73, 76, 80, 83, 89, 97-8, 110, 121, 126, 133, 134, 136
rosemary 20, **54**, 77, 83, 90, 97, 115, 119, 123, 125, 133, 139
Rosmarinus officinalis see rosemary

INDEX

roundworm 62, 138
Rubus idaeus see raspberry
Russian Comfrey 34
ruta grav. 133
rutin 29, 42, 95

sage **55**, 108, 134
St John's Wort *see* hypericum
salicylic acid 47, 61
Salix alba see willow
salt water 79
Salva officinalis see sage
Sandberry *see* uva-ursi
sarcoids **126-7**
scabies 26
schooling 96, 114
scouring 47, 53, 56-7, 68, 70, 84, 117, 118, 119, **127-8**
  foal scour **92**
scratches *see* mud fever
Scratch-weed *see* clivers
scullcap **55-6**, 88, 90
scurfy skin 29, 76
Scutch *see* couch grass
Scutellaria laterifolia see scullcap
seasons, irregular or over-lengthy 98
seaweed 97, 113, 147
  see also kelp
Seawrack *see* kelp
sedative action 52, 65, 88
sedatives, conventional 88
seeds 138
selenium 110
septicaemia 70
sheaths
  cleaning 58-9, **128-9**
  infected 58, 59
shock 27, 88, 103, **129**, 143
Short Buchu *see* buchu
Siberian ginseng 91, 101
silica 31, 32, 35, 77, 97
Silybum marianum see milk thistle

silymarin 48, 106
Single Chamomile *see* chamomile, German
skin condition *see* coat and skin condition
skin irritation 30, 44, 46, 58
  see also dermatitis; eczema; greasy heels; mud fever; rain scald; sweet itch
Skullcap *see* scullcap
slippery elm 20, **56-7**, 79, 85, 92, 113, 118, 119, 128, 135, 140
smegma, overproduction of 58, 129
sodium 49, 110
soft swellings 31, 43, 82, **130-1**
soil, wood and droppings, eating 109
Soldiers' Woundwort *see* yarrow
soles, bruised **76**
Solidago virgaurea see golden rod
sores 29, 57, 61
Spanish Liquorice *see* liquorice
Spearmint *see* mint
Spike Lavender *see* lavender
spinal impingement or pain 96
spinal misalignment 82, 96, 98
staling 28
stallions at stud 129
Staunchweed *see* yarrow
steroidal saponins 39
Stinging Nettle *see* nettle
stings 40
stomach cramps 59
strangles 37
stress 31, 55, 59, 74, 78-9, 84, 90, 93, 96, 103, 114, 135, 137, 143

stupor 88
sulphur 30, 40, 110, 113, 131, 132
summer dermatitis/eczema *see* sweet itch
sweating 80, 84
Sweet Elm *see* slippery elm
sweet itch 10, 64, **131-2**
Symphytum spp. *see* comfrey

Taraxacum officinale see dandelion
tea tree oil **57**, 112, 113, 125, 133, 140
tendon and ligament strain/damage **132-3**
tension 31, 55, 90, 115
thiamine 53, 70, 110
Thoroughwort *see* boneset
threadworm 62, 138
throats, inflamed 51
thrush **133-4**
thuja 127, 136
thyme 20, **58**, 70, 113, 115, 123
thymol 54, 58
Thymus vulgaris see thyme
thyroid function 44, 73, 93
thyroid stimulating hormone (TSH) 93
tick and tick bites 40, 57
tonics 30, 48, 49, 50, 51, 52, 53, 80, 89, 91, 101, 105
  see also digestive tonics
tooth problems 96, 119
toxin elimination 82, 83, 86, 100, 102, 115, 122, 123, 126
Trifolium pratense see red clover
Trigonella foenum-graecum see fenugreek
tumours 52, 73, 79
  see also sarcoids
turmeric 21, 43
Twitch Grass *see* couch grass

175

INDEX

ulcers 30, 38, 39, 56, 61
  see also gastric ulcers; mouth ulcers
Ulmus rubra, U. fulva see slippery elm
urethritis 28, 37
urinary antiseptics 28, 30, 35, 58, 81
urinary infections 28, 30, 31, 37, 41, 43, 46, 50, 58, 63
  see also cystitis; urethritis
urinary system 30, 101
Urtica dioica see nettle
uterine muscles 36, 51, 90, 91
uterus 52, 90
  infected 58
uva-ursi **58-9**, 81, 129

vaccines, allergic reaction to 130
valerian 55, **59-60**, 75, 79, 86, 88, 90, 93, 96, 117, 135, 137
Valeriana officinalis see valerian
vasodilatory action 28, 29, 42, 83, 102, 104, 117
Verbena officinalis see vervain
vervain 20, **60**, 75, 88, 96, 100, 105, 110
veterinary care and expertise 13, 72
viral infections 37, 40, 52, 136
Virginian Skullcap *see* scullcap
vitamin A 35, 39, 53, 70, 110
vitamin B 35, 39, 66, 88, 97, 100, 110, 134
vitamin B12 32, 66, 73, 110, 134
vitamin C 35, 39, 49, 50, 53, 70, 73, 76, 80, 83, 89, 97, 98, 110, 136
vitamin D 35, 97, 110
vitamin E 39, 91, 97, 110, 127, 137
vitamin K 53, 66, 70, 110
vitamin deficiency *see* mineral/vitamin deficiency
Vitex agnus-castus see agnus castus
vulnerary herbs 20

walnut remedy 144
warts 36, **136-7**
wash-down lotions 27, 61
water quality **67-8**, 77
watercress 110
Watermint *see* mint
weaving **137**
weight loss 80, 117

White Horehound *see* horehound
White Willow *see* willow
Whitethorn *see* hawthorn
Wild Briar *see* rosehip
willow **61**, 83, 102, 110
windgalls/windpuffs 31, 43, 130
witch hazel 27, 38, **61**, 76, 86, 139
Witchgrass *see* couch grass
wolf teeth 96
worms 40, 43, 48, 62, 78, 92, **138-9**
wormwood 17, **62**, 80, 85, 88, 93, 127, 136, 138
wounds 27, 61, **139-40**
  bathing 41, 44, 70, 139
  healing 29, 32-3, 33, 37, 56, 63, 103
  infected 41, 57, 70
  puncture wounds 20, 140
Woundwort *see* golden rod

yarrow 20, 28, **63**, 76, 81, 87, 93, 107, 110, 115, 123, 139
yoghurt 57, **68**, 70, 81, 117, 128, 135

zinc 110